Studies in Classification, Data Analysis, and Knowledge Organization

Studies in Classification, Data Analysis, and Knowledge Organization is a book series which offers constant and up-to-date information on the most recent developments and methods in the fields of statistical data analysis, exploratory statistics, classification and clustering, handling of information and ordering of knowledge. It covers a broad scope of theoretical, methodological as well as application-oriented articles, surveys and discussions from an international authorship and includes fields like computational statistics, pattern recognition, biological taxonomy, DNA and genome analysis, marketing, finance and other areas in economics, databases and the internet. A major purpose is to show the intimate interplay between various, seemingly unrelated domains and to foster the cooperation between mathematicians, statisticians, computer scientists and practitioners by offering well-based and innovative solutions to urgent problems of practice.

More information about this series at http://www.springer.com/series/1564

Simona Balzano · Giovanni C. Porzio ·
Renato Salvatore · Domenico Vistocco ·
Maurizio Vichi
Editors

Statistical Learning and Modeling in Data Analysis

Methods and Applications

Springer

Editors
Simona Balzano
Department of Economics and Law
University of Cassino and Southern Lazio
Cassino, Italy

Giovanni C. Porzio
Department of Economics and Law
University of Cassino and Southern Lazio
Cassino, Italy

Renato Salvatore
Department of Economics and Law
University of Cassino and Southern Lazio
Cassino, Italy

Domenico Vistocco
Department of Political Science
University of Naples Federico II
Naples, Italy

Maurizio Vichi
Department of Statistical Sciences
Sapienza University of Rome
Rome, Italy

ISSN 1431-8814 ISSN 2198-3321 (electronic)
Studies in Classification, Data Analysis, and Knowledge Organization
ISBN 978-3-030-69943-7 ISBN 978-3-030-69944-4 (eBook)
https://doi.org/10.1007/978-3-030-69944-4

Mathematics Subject Classification: 62-06, 62-07, 62Fxx, 62Gxx, 62Hxx, 62Jxx, 62Kxx

This Springer imprint is published by the registered company Springer Nature Switzerland AG
The registered company address is: Gewerbestrasse 11, 6330 Cham, Switzerland

Preface

This book offers a collection of papers focusing on methods for statistical learning and modeling in data analysis. A series of interesting applications are offered as well. Several research topics are covered, ranging from statistical inference and modeling to clustering and factorial methods, from directional data analysis to time series analysis and small area estimation. Applications deal with new analyses within a variety of fields of interest: medicine, finance, engineering, marketing, cyber risk, to cite a few.

The book arises as post-proceedings of the 12th meeting of the CLAssification and Data Analysis Group (CLADAG) of the Italian Statistical Society (SIS), held in Cassino (IT), on September 11–13, 2019. The first CLADAG meeting was held in 1997, in Pescara (IT). CLADAG is also a member of the International Federation of Classification Societies (IFCS), founded in 1985. CLADAG promotes advanced methodological research in multivariate statistics with a special vocation towards Data Analysis and Classification. It supports the interchange of ideas in these fields of research, including the dissemination of concepts, numerical methods, algorithms, computational and applied results. This book is thus in line with the main CLADAG goals.

Thanks to the participation of renowned speakers, coming from 28 different countries, the scientific program of the CLADAG 2019 Conference was particularly engaging. It saw 5 Keynote Lectures, 32 Invited Sessions, 16 Contributed Sessions, a Round Table, and a Data Competition. The richness of the Conference program, and hence of this book, is definitely due to the Conference Scientific Committee and particularly to its Chair Francesca Greselin. We are indebted to their work. We are also indebted to the anonymous referees. They did a great job and helped us to improve the overall quality of this book.

Our gratitude also goes to the staff of the Department of Economics and Law, University of Cassino and Southern Lazio, who supported the conference and contributed to its success. A special thank goes to Livia Iannucci, who worked side by side with the Local Organizing Committee offering her precious administrative support before, during, and after the conference.

Above all, we are thankful to all the participants and to those who, among them, have chosen this book to share their research findings. Our wish is that this book will contribute to foster the creation of new knowledge in the field.

Cassino, Italy
26 November 2020

Simona Balzano
Giovanni C. Porzio
Renato Salvatore
Domenico Vistocco
Maurizio Vichi

Contents

Interpreting Effects in Generalized Linear Modeling

Alan Agresti, Claudia Tarantola, and Roberta Varriale

Abstract With nonlinear link functions in generalized linear models, it can be difficult for nonstatisticians to understand how to interpret the estimated effects. For this purpose, it can be helpful to report approximate effects based on differences and ratios for the mean response. We illustrate with effect measures for models for categorical data. We mainly focus on binary response variables, showing how such measures can be simpler to interpret than logistic and probit regression model parameters and their corresponding effect measures, such as odds ratios. For describing the effect of an explanatory variable on a binary response while adjusting for others, it is sometimes possible to employ the identity and log link functions to generate simple effect measures. When such link functions are inappropriate, one can still construct analogous effect measures from standard models such as logistic regression. We also summarize recent literature on such effect measures for models for ordinal response variables. We illustrate the measures for two examples and show how to implement them with R software.

Keywords Binary data · Link functions · Logistic regression · Ordinal data · Partial effects · Probit models

A. Agresti (✉)
Department of Statistics, University of Florida, Gainesville, USA
e-mail: agresti@ufl.edu

C. Tarantola
Department of Economics and Management, University of Pavia, Pavia, Italy
e-mail: claudia.tarantola@unipv.it

R. Varriale
Istat, Rome, Italy
e-mail: varriale@istat.it

S. Balzano et al. (eds.), *Statistical Learning and Modeling in Data Analysis*,
Studies in Classification, Data Analysis, and Knowledge Organization,
https://doi.org/10.1007/978-3-030-69944-4_1

1 Introduction

For n independent response observations $\{y_i\}$ with $\{\mu_i = E(y_i)\}$ and p explanatory variables, consider the generalized linear model (GLM) using link function g,

$$g(\mu_i) = \alpha + \beta_1 x_{i1} + \cdots + \beta_p x_{ip}.$$

Except for the normal linear model, standard GLMs use nonlinear link functions such as logistic and probit models for binary data and loglinear models for count data. For many such link functions, interpretation of numerical values of the estimates $\{\hat{\beta}_k\}$ can be difficult, especially for nonstatisticians and for methodologists who are mainly familiar with ordinary linear models.

In this paper, we show how to report simple approximations of the effects, such as based on a linearization of the model, that can be much simpler to interpret. The approximations can be expressed in terms of differences and ratios for the mean of the response variable. We illustrate with effect measures for popular models for categorical data. Section 2 focuses on models for binary regression models, using such measures to supplement standard effects such as the odds ratio. Section 3 summarizes recent literature on effect measures for models for ordinal response variables that apply link functions to cumulative probabilities. Implementation of the measures is simple using `R` software. We illustrate effect interpretation for binary data with an Italian study to model an employment response variable and for ordinal data with a study of mental impairment.

2 Interpreting Effects in Generalized Linear Models for Binary Data

Models for binary responses that apply nonlinear link functions to the probability of "success," such as logistic and probit regression models, have model effect parameters that are not as simple to interpret as slopes and correlations for ordinary linear regression. For example, logistic regression has effects most naturally interpreted using odds ratios. To compare two levels of an explanatory variable such as two groups, however, it is easier for practitioners to understand a difference or a ratio of *probabilities* than a ratio of *odds*. In practice, even some statisticians misinterpret the odds ratio as if it were a ratio of probabilities. When two groups have probabilities close to 0, the ratio of odds is similar to the ratio of probabilities, but this is not true otherwise. For example, when the probabilities exceed 0.2, the odds ratio is better approximated by the *square* of the ratio of probabilities [7]. If an odds ratio is 9, one group may have success probability merely about 3 times the success probability for the other group.

Another aspect of logistic and probit regression that is due to the nonlinear link function is the dependence of a variable's effect on the other explanatory variables in the model, even when those variables are uncorrelated with the one of interest. Suppose explanatory variables x_1 and x_2 are uncorrelated, such as in many experimental designs. In ordinary linear models, the estimated effect of x_1 is the same when x_1 is the sole predictor as when x_1 and x_2 are joint predictors. For logistic regression, this is not the case with model-based odds effect measures. For instance, the effect β_1^* when x_1 is the sole predictor relates to the effect β_1 when x_2 is also in the model by $\beta_1^* \approx \beta_1\sqrt{3.29/[3.29 + \beta_2^2 \text{var}(x_2)]}$, where $3.29 = \pi^2/3$ is the variance of the standard logistic distribution [6]. For the model with probit link, $\beta_1^* = \beta_1\sqrt{1/[1 + \beta_2^2 \text{var}(x_2)]}$. Equality of the effects in the two cases is, however, approximately true for the simpler measures discussed next.

2.1 *Alternatives to the Logit and Probit Links with Binary Responses*

For binary responses, the logit and probit link functions are used almost exclusively. Sometimes, however, we can also use the log and the identity link functions.

- The identity link provides similar fits as the logit or probit link when $P(y = 1)$ falls mainly between about 0.2 and 0.8. It has simpler interpretations, as the model parameters relate to *differences of probabilities* instead of *ratios of odds*.
- The log link provides similar fits as the logit or probit link when $P(y = 1)$ falls mainly below 0.25, and similar to those models with log link applied to $P(y = 0)$ when probabilities are uniformly above 0.75. It has simpler interpretations, as the model parameters relate to *ratios of probabilities* instead of *ratios of odds*.
- With uncorrelated explanatory variables, the effects with log and identity links are the same in the full model as in marginal models with sole predictors, which is not true with logit or probit links.

We illustrate the first two points with data from Istituto Nazionale di Statistica (Istat), the Italian government agency for official statistics. We fitted models to some data from a simple random sample of 100,000 Italians from the Toscana region of Italy in December 2015. For the binary response y = whether employed (where $y = 1$ means that the person is present in some administrative source about labor statistics), we use explanatory variables x_1 = gender (1 = female, 0 = male), x_2 = whether an Italian citizen (1 = yes, 0 = no), and x_3 = whether receiving a pension (1 = yes, 0 = no).

Consider first the 27,775 subjects in the survey having age over 65. For the 8 combinations of x_1, x_2, x_3, the sample proportions employed fall between 0.02 and 0.12. The main effects logit and log-link model fits are

$$\text{logit}[\hat{P}(y = 1)] = -1.869 - 1.324x_1 - 0.429x_2 + 0.216x_3,$$

$$\log[\hat{P}(y = 1)] = -2.037 - 1.239x_1 - 0.362x_2 + 0.200x_3.$$

The absolute difference in fitted proportions, averaged over the 27,775 cases, is 0.0001. For the log-link model, the exponentiated coefficients estimate probability ratios; for example, adjusting for x_2 and x_3, the probability that a woman is employed is estimated to be $\exp(-1.239) = 0.290$ times the probability that a man is employed.

Consider next the 72,225 subjects having age under 65. For the 8 combinations of x_1, x_2, x_3, the sample proportions employed fall between 0.18 and 0.74. The main-effects logit and identity-link model fits are

$$\text{logit}[\hat{P}(y = 1)] = 0.350 - 0.644x_1 + 0.702x_2 - 1.874x_3,$$

$$\hat{P}(y = 1) = 0.587 - 0.139x_1 + 0.151x_2 - 0.408x_3.$$

The absolute difference in fitted proportions, averaged over the 72,225 cases, is only 0.004. For the identity-link model, the coefficients estimate differences of probabilities. For instance, adjusting for x_2 and x_3, the probability that a woman is employed is estimated to be 0.139 lower than the probability that a man is employed.

Of course, an advantage of using a standard binary model such as logistic regression is that it is relevant regardless of the range of values for $P(y = 1)$. However, when the probabilities are in the appropriate ranges, we believe that the log-link model and identity-link model can supplement the logit-link model by providing effect interpretations that are simpler for many to understand. For further details, including the data for this example, see [4].

2.2 Probability Effects Measures for Logistic Models

When we fit standard binary models such as logistic or probit regression, summary measures based on differences and ratios of probabilities can summarize the size of the effects. Such effects, which are available even when we do not consider separate models with identity or log link functions, also exhibit greater stability in terms of the impact of uncorrelated explanatory variables.

A simple summary for the effect of an explanatory variable x_k averages the rate of change in $P(y = 1)$, as a function of x_k. For this, we express the model as

$$F^{-1}[P(y = 1)] = \alpha + \beta_1 x_1 + \cdots + \beta_p x_p, \tag{1}$$

where the link function $g = F^{-1}$ is the inverse of a standard cdf. For logistic regression, $F(z) = \exp(z)/[1 + \exp(z)]$ is the standard logistic cdf. For probit regression, F is the standard normal cdf, which we denote by Φ. Let $f(y) = \partial F(y)/\partial y$ denote the corresponding probability density function. For a quantitative explanatory variable x_k, the rate of change in $P(y = 1)$ when other explanatory variables are fixed at certain values (i.e., the partial effect) is

$$\partial P(y = 1|\boldsymbol{x})/\partial x_k = f(\alpha + \beta z + \beta_1 x_1 + \cdots + \beta_p x_p)\beta_k.$$

For the logit link, the partial effect for x_k on $P(y = 1)$ has the expression

$$\partial P(y = 1|\boldsymbol{x})/\partial x_k = \beta_k P(y = 1|\boldsymbol{x})[1 - P(y = 1|\boldsymbol{x})].$$

This takes values bounded above by its highest value of $\beta_k/4$ that occurs when $P(y = 1|\boldsymbol{x}) = 1/2$. For probit models, the highest value of this instantaneous rate of change is $\beta_k/\sqrt{2\pi}$, also when $P(y = 1|\boldsymbol{x}) = 1/2$. These maximum values need not be relevant, as $P(y = 1)$ need not be near 1/2 for most or all the data.

Any particular way of fixing values of the explanatory variables has its corresponding partial effect value for x_k. The authors of [5] summarize various versions. For example, the *average partial effect* (called *average marginal effect* in some literature and software) estimates the partial effect of x_k at each of the n sample values of the explanatory variables, and then averages them. For a categorical explanatory variable, one would instead use a *discrete change*, estimating the change in $P(Y = 1)$ for a change in an indicator variable. For comparing two groups having indicator variable z, for instance, for the n sample observations, we could find the difference between estimates of $P(y = 1)$ when $z = 1$ and when $z = 0$ at the sample values for the other predictors, and average the obtained differences. Discrete changes are also relevant for quantitative explanatory variables, to summarize estimated changes in $P(y = 1)$ over a particular range of x_k values. For example, to summarize the effect of a quantitative variable x_k on y, it can be useful to report the difference between the model-fitted estimate of $P(y = 1)$ at the maximum and minimum values of x_k, when other explanatory variables are set at particular values such as their means.

A measure that we've not seen proposed for the two-group comparison focuses on average partial *ratios* of estimated probabilities for the groups. For example, we could average the n ratios of probability estimates, or average the n log ratios of probability estimates and then exponentiate that average. The authors of [4] discuss such measures, which are useful when fitted probabilities are near 0 for the groups being compared.

The effect measures are available in R software. For instance, here is how to use an existing R package to estimate average partial effects for the younger sample of the Istat data, after fitting a logistic regression model

```
-------------------------------------------------------------------
mod.logit <- glm(y ~ x1 + x2 + x3, family=binomial, data=younger)
library(mfx) # library with functions for rate of change effects
logitmfx(mod.logit, atmean=FALSE, data=younger)
Marginal Effects:
              dF/dx     Std. Err.
x1       -0.1406203    0.0034589 # average estimated diff. of P(y=1)
x2        0.1582009    0.0051184 # for binary predictors
x3       -0.4160246    0.0050788
-------------------------------------------------------------------
```

The reported discrete change effect for x_1 indicates that with the logit link, the average difference in the estimated probability of employment between women and men is -0.141. This is close to the estimated gender effect of -0.139 shown above in using the identity link. For the older sample, an R function presented by [4] estimates an average ratio of employment probabilities of 0.2895 for comparing women with men. This is close to the estimated gender effect of $\exp(-1.239) = 0.2897$ using the log link.

3 Interpreting Effects in Models for Ordinal Responses

Summary effect measures are also relevant in GLMs for multi-category response variables. Here, we consider a standard model for ordinal responses. For observation i on ordinal response variable y having c categories, the *cumulative link model* is

$$\text{link}[P(y_i \leq j)] = \alpha_j + \sum_k \beta_k x_{ik}, \quad j = 1, ..., c-1, \tag{2}$$

for link functions such as the logit and probit. With the logit link, effects now refer to odds ratios that relate to outcome below instead of above any particular point on the response scale. For the probit link, effects are especially difficult to interpret, as β_k is the change in $\Phi^{-1}[P(y_i \leq j)]$ for each 1-unit increase in x_k, adjusting for the other explanatory variables.

A common approach to aid in interpretation focuses on means and probabilities for an underlying latent variable model that yields this ordinal-response model. For a latent variable y^*, suppose that $y_i^* = \sum_j \beta_j x_{ij} + \epsilon_i$, where ϵ_i has cdf G with mean 0, and suppose thresholds $-\infty = \alpha_0 < \alpha_1 < \ldots < \alpha_c = \infty$ exist such that $y_i = j$ if $\alpha_{j-1} < y_i^* \leq \alpha_j$. Then, it follows (e.g., see [1], pp. 303–304) that $G^{-1}[P(y_i \leq j)] = \alpha_j - \sum_j \beta_j x_{ij}$. The effects in the ordinal model as parameterized in equation (2) are merely the negatives of those in the latent variable model. So, for instance, the cumulative link model with probit link is valid when an ordinary normal linear model holds for the underlying response, and β_k in the ordinal model has the interpretation that a 1-unit increase in x_k corresponds to a change in $E(y^*)$ of β_k standard deviations, adjusting for the other explanatory variables. But this interpretation can be rather obscure for non-methodologists. Alternative summaries can focus on the probability scale, including generalizations of the measures we've discussed for binary data.

The authors of [2] suggested a simple probability summary to compare two groups, adjusting for the other p explanatory variables: At a setting $(x_1, \ldots, x_p)$ of explanatory variables, let y_1^* and y_2^* denote independent latent variables when a group indicator is $z = 1$ and when $z = 0$, respectively. Consider $P(y_1^* > y_2^*)$. For the latent variable model that generates the cumulative link model with probit link, with parameter β as the coefficient of an indicator variable for the two groups, they showed that $P(y_1^* > y_2^*) = \Phi(\beta/\sqrt{2})$ at any setting of the explanatory variables. For the cumulative link model with logit link, they showed that $P(y_1^* > y_2^*) \approx \exp(\beta/\sqrt{2})/[1 + \exp(\beta/\sqrt{2})]$.

However, the latent variable model is not always appropriate. The authors of [3] surveyed ways to use average partial effect measures to summarize effects for ordinal models for the observed data, without reference to latent variables. As any x_k increases, cumulative link models that contain solely main effects imply monotonicity in the extreme outcome category (1 and c) probabilities. One can summarize the effect by the average rates of change

$$\frac{1}{n}\sum_{i=1}^{n} \partial P(y_i = 1|\boldsymbol{x}_i)/\partial x_k$$

$$\frac{1}{n}\sum_{i=1}^{n} \partial P(y_i = c|\boldsymbol{x}_i)/\partial x_k.$$

The authors of [3] prepared `R` functions for such effects.

We illustrate with an example from [3], for which the data come from a study of effects associated with mental health. The response variable was an ordinal measure of mental impairment, with categories (1 = well, 2 = mild impairment, 3 = moderate impairment, 4 = impaired). The explanatory variables were x_1 = socioeconomic status (SES: 1 = high, 0 = low) and x_2 = a life events index that is a numerical composite measure of the number and severity of important life events such as birth of child, new job, divorce, or death in family that occurred to the subject within the past three years. The life-events index takes values on the nonnegative integers between 0 and 9, with mean 4.3 and standard deviation 2.7. The $n = 40$ observations are available as shown in the following `R` output. Here, we use the `R` functions from [3] to summarize effects on the extreme categories (1 = well, 4 = impaired) for the fit of the cumulative link model with logit link and main effects of the explanatory variables

```
-------------------------------------------------------------------
> Ord<-read.table("http://www.stat.ufl.edu/~aa/glm/data/Mental.dat",
+                     header=TRUE)
> library(MASS) # for polr = proportional odds logistic regression
> fit.logit <- polr(y ~ ses + life, method="logistic", data=Ord)
> summary(fit.logit)
        Value  Std. Error  t value
ses   -1.1112      0.6109   -1.819
life   0.3189      0.1210    2.635
> ocAME(fit)             # ordinal "average marginal effect" function
$ME.1                    # WELL category for mental impairment
      effect  std.error
ses    0.198      0.104
life  -0.057      0.019
$ME.4                    # IMPAIRED category for mental impairment
      effect  std.error
ses   -0.171      0.094
life   0.048      0.017
-------------------------------------------------------------------
```

At the observed values for life events and SES, the rate of change in the estimated probability per unit change in life events averages to −0.057 for the *well* outcome and to 0.048 for the *impaired* outcome. At the observed values for life events, when SES increases from 0 to 1, the estimated probability of the *well* outcome increases by an average of 0.198 and the estimated probability of the *impaired* outcome decreases by an average of 0.171.

4 Future Potential Research

This article has focused on generalized linear models, with emphasis on models for categorical response variables, but the need for interpretable measures is even greater for more complex models. Future research could develop and apply simple summary measures for other models.

The authors of [4] showed how to apply average partial effect measures to *generalized additive models* for binary data, which replace the linear predictor by additive unspecified smooth functions. The measures presented there are appropriate when relationships are monotone, but often that is not the case when such models are used. Even when it is the case, difference or ratio effects are sometimes highly variable across the range of an explanatory variable, and a single summary may be too simplistic.

For categorical responses, using alternative link functions to aid in interpretation would be useful for marginal models, whether fitted by GEE methods or maximum likelihood. The binary and log links are more challenging for random effects models, as the usual assumption of normally-distributed random effects adds another restriction to models with bounded range values. However, in either case, it should be possible to generalize the average partial effect measures.

References

1. Agresti, A.: Categorical Data Analysis, 3rd edn. Wiley, Hoboken (2013)
2. Agresti, A., Kateri, M.: Ordinal probability effect measures for group comparsons in multinomial cumulative link models. Biometrics **73**, 214–219 (2017)
3. Agresti, A., Tarantola, C.: Simple ways to interpret effects in modeling ordinal categorical data. Stat. Neerlandica **72**, 210–223 (2018)
4. Agresti, A., Tarantola, C., Varriale, R.: Simple ways to interpret effects in modeling binary responses. Submitted for Trends and Challenges in Categorical Data Analysis (ed. M. Kateri and I. Moustaki), to be published by Springer (2021)
5. Long, J.S., Mustillo, S.A.: Using predictions and marginal effects to compare groups in regression models for binary outcomes. Sociol. Methods Res (2018). https://doi.org/10.1177/0049124118799374
6. Mood, C.: Logistic regression: why we cannot do what we think we can do, and what we can do about it. Europ. Sociol. Rev. **26**, 67–82 (2010)
7. VanderWeele, T.J.: On the square-root transformation of the odds ratio for a common outcome. Epidemiology **28**, e58–e60 (letter to the editor) (2017)

ACE, AVAS and Robust Data Transformations

Anthony C. Atkinson, Marco Riani, Aldo Corbellini, and Gianluca Morelli

Abstract Unlike the Box-Cox transformation, that of Yeo and Johnson for the response of a linear model can be applied when the observations are not constrained to be positive. We study the extended Yeo–Johnson transformation in which positive and negative observations can be transformed with different parameter values. The procedure is illustrated for data with many outliers. The data are cleaned with a robust method, the forward search, and the obtained transformations compared with the results from two nonparametric transformation methods based on data smoothing.

Keywords Box-Cox transformation · Extended Yeo–Johnson transformation · Fan plot · Forward search · Nonparametric transformation · Smoothing

1 Introduction

The widely used parametric family of power transformations introduced by [3] is only applicable to positive observations. Yeo and Johnson [7] spliced together two Box-Cox transformations to provide a one-parameter family of transformations for data that can be positive or negative. Atkinson et al. [2] extended the Yeo–Johnson

A. C. Atkinson
Department of Statistics, London School of Economics, London, UK
e-mail: a.c.atkinson@lse.ac.uk

M. Riani (✉) · A. Corbellini · G. Morelli
Dipartimento di Scienze Economiche e Aziendale and Interdepartmental Centre for Robust Statistics, Università di Parma, Parma, Italy
e-mail: mriani@unipr.it

A. Corbellini
e-mail: aldo.corbellini@unipr.it

G. Morelli
e-mail: gianluca.morelli@unipr.it

S. Balzano et al. (eds.), *Statistical Learning and Modeling in Data Analysis*, Studies in Classification, Data Analysis, and Knowledge Organization,
https://doi.org/10.1007/978-3-030-69944-4_2

transformation to allow different parameter values for the transformation of positive and negative observations; they illustrate the usefulness of this procedure through the analysis of two sets of data.

Nonparametric transformations provide an alternative to such families of parametric transformations. The purpose of this short paper is to compare parametric and nonparametric transformations for a set of data that contain outliers, to illustrate a robust method for cleaning the data of outliers and to compare transformations on the cleaned data.

2 Extended Parametric Transformations

The purpose of these parametric transformations is to achieve a response which is approximately normally distributed with errors of constant variance and a linear model of simple form. For comparisons of estimates of parameters for different values of λ, many authors, starting with [3], stress the importance of working with a normalized transformation allowing for the change of scale of the observations with transformation. For the Box-Cox transformation, the normalized transformation is

$$z(\lambda) = (y^\lambda - 1)(z^\lambda - 1)/(\lambda \dot{y}^{\lambda-1}) \; (\lambda \neq 0); \; \dot{y} \log y \; (\lambda = 0), \tag{1}$$

where $\dot{y}$ is the geometric mean of y and J, the Jacobian of the transformation is given by $\log J = n(\lambda - 1) \log \dot{y}$. The linear model to be fitted is $z(\lambda) = X\beta(\lambda) + \varepsilon$, where X is $n \times p$, β is a $p \times 1$ vector of unknown parameters and the variance of ε is σ^2.

The normalized transformation for the two-parameter extended Yeo–Johnson (EYJ) transformation is given by [2]. There are now four regions of y, rather than two, with distinct forms of $z(\lambda)$ and the Jacobian is now a more complicated function of the observations.

3 Robustness and the Fan Plot

We use a robust procedure, the Forward Search [1] to order the data by closeness to the fitted model. The procedure starts from a carefully chosen subset of $m_0 = p + 1$ observations and moves forward increasing the subset size m by introducing the observation, not used in fitting, that is closest to the fitted model, until all observations have been fitted. Outliers, if any, enter at the end of the search.

Outliers in one value of λ may not be so for some other values. We, therefore, need to repeat the forward search for a grid of values of λ. For each resultant ordering of the data, we monitor evidence for the correctness of the transformation as m increases. We include the constructed variable $w(\lambda) = \partial z(\lambda)/\partial\lambda$ in the linear model for the EYJ transformation. The approximate score statistic for the value λ_0 is the t-test for the significance of $w(\lambda_0)$ in the regression. The plot of trajectories of the score statistic against subset size for a set of values of λ is called a fan plot.

Constructed variables for the one-parameter Yeo–Johnson transformation are given by [2]. They further derive constructed variables for testing whether positive and negative observations require the same transformation. These come from the extended transformation in which one kind of response has the parameter $\lambda + \alpha$ and the other λ. The test is for $\alpha = 0$.

4 Augmented Investment Fund Data

As our example, we analyze data on the relationship between the medium term performance of 309 investment funds and two indicators. Of these funds, 99 have negative performance. To examine the properties of transformation procedures in the presence of outliers, we augmented the data with 40 outliers, to produce a data set in which the outliers are evident after transformation, but not before. The analysis of the uncontaminated data [2] concludes that the negative observations need transformation with parameter $\lambda_N = 0$, which for the EYJ is not the log transformation. The positive observations need no transformation ($\lambda_P = 1$). The data are well behaved, with no evidence of any outliers.

The fan plot for the augmented data indicates that the majority of the outliers enters the subset at the end of the search; the structure of the plot changes for $m > 310$. The extended fan plot with separate trajectories for positive and negative observations shows that different transformations are required for the two parts of the data. The best values are $\lambda_N = 0$ and $\lambda_P = 1$ when the trajectories of the score statistics for positive and negative observations are similar to those for the overall data until m is around 320. This is the transformation found for the uncontaminated data.

5 Robust Analysis

We now identify the outlying observations by a forward search analysis of the data with the recommended transformation $\lambda_P = 1$ and $\lambda_N = 0$.

The left-hand panel of Fig. 1 shows a forward plot of all 349 scaled residuals of the augmented data for a wide range of values of m. There is an upper band of residuals, in blue in the online version of the paper, separated from a lower band of 37 residuals, shown in red. What is remarkable is the stability of this pattern, until $m = 310$, indicative of a set of data without outliers and with normally distributed errors and the second group of observations, not included in the subsets used for fitting.

The highlighted, red, residuals were identified by brushing the plot. That is, we selected all the trajectories that lie within the brush in the centre of the figure. The right-hand panel of the figure shows a linked forward plot of minimum Mahalanobis

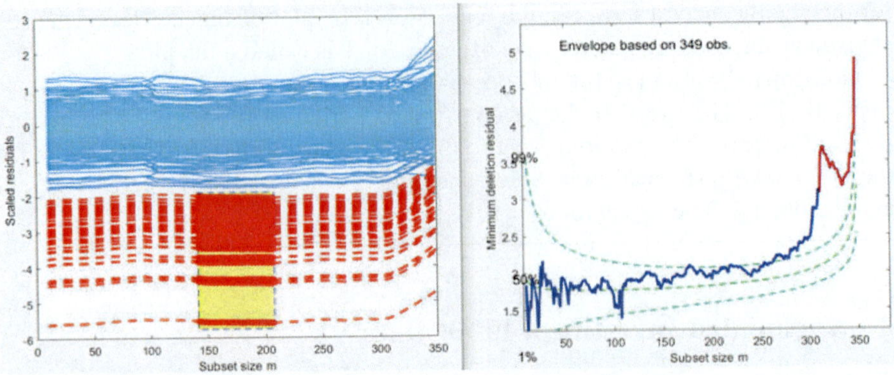

Fig. 1 Augmented investment fund data; brushing linked plots from the forward search when $\lambda_P = 1$ and $\lambda_N = 0$. Left-hand panel, trajectories of residuals from the forward search with the residuals from 37 observations highlighted by brushing. Right-hand panel, linked forward plot of minimum deletion residuals during the search with the 37 brushed values shown in red

distances, with the trajectory of the 37 brushed observations shown in red. These are indeed the last observations to enter the search. Our automatic procedure for outlier detection [5] in fact identifies 35 outliers.

6 Nonparametric Transformations

It is clear from the results of the previous sections that the contaminated data need both cleaning and transformation. The purpose of this section is to determine what information nonparametric transformations provide on the presence of outliers, the transformation of the data and whether the parametric extended Yeo–Johnson transformation can be improved by further transformation. The parametric transformations produce a smooth relationship between $z(\lambda)$ and the original y. A nonparametric alternative is to use smoothing to estimate this relationship. We use two such methods, ACE—Alternating Conditional Expectations—[4] and AVAS [6] in which the transformation for the response is intended to yield additivity and variance stabilization. We consider only response transformation, comparing models through the value of unadjusted R^2.

We start with the extended Yeo–Johnson transformation, using the parameter estimates $\lambda_P = 1$ and $\lambda_N = 0$ for all comparisons. First we look at the contaminated data before and after cleaning. The left-hand panel of Fig. 2 shows the QQ plot of the residuals of all 349 observations from regression with the original response. The sigmoid shape of this plot indicates that the observations are not normally distributed. The right-hand panel is the QQ plot for residuals of the transformed cleaned data. The distribution of residuals is much closer to normality, although the centre of the curve indicates that many small residuals are slightly too large in absolute value.

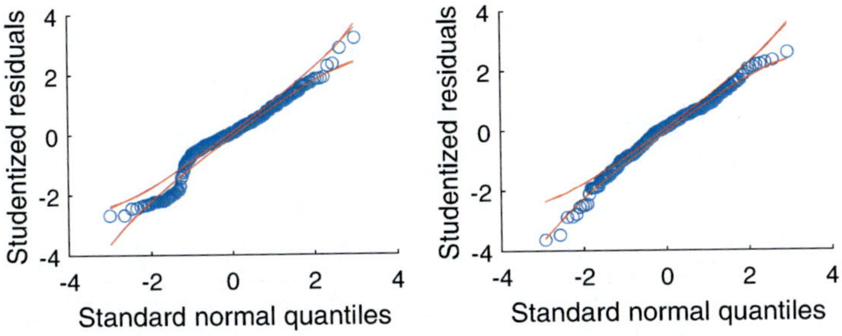

Fig. 2 Comparison of normal QQ plots of residuals. Left-hand panel, untransformed contaminated data. Right-hand panel, transformed cleaned data

Table 1 Investment fund data: summary properties of regression for parametric and nonparametric transformations of contaminated and cleaned data

	Contaminated	Cleaned and transformed
Untransformed	0.399	–
EYJ	0.356	0.783
AVAS	0.241	0.778
ACE	0.421	0.806
ACE (monotonic)	0.417	0.805

The value of R^2 for regression on the untransformed contaminated data is 0.399. For the cleaned transformed data it is 0.783 and for the uncontaminated data 0.816. The left-hand column of Table 1 lists the values of R^2 achieved by regression on parametric and nonparametric transformations of the contaminated data. The largest value is 0.421 for unconstrained ACE. The monotonicity constraint on ACE comes from isotonic regression on the unconstrained transformation and yields a slightly reduced value of 0.417. AVAS produces a value of 0.241, less than that for EYJ. Figure 3 provides plots of transformed against untransformed response for these four transformations.

The top left-hand panel of the figure shows that the EYJ transformation for $y > 0$ is linear (no transformation), whereas for negative y, the transformation is concave, transforming the more negative observations to be more extreme. AVAS, in the top right-hand panel, provides a more smooth concave curve, which not only makes the more negative values more extreme but makes the more positive values less extreme. Unconstrained ACE is virtually linear for $y > 12$, but shrinks in the most negative observations, some of which are outliers. Constrained ACE is formed by isotonic regression on the unconstrained version, and as the figure shows, is similar in structure to ACE. Both transformations show several points of inflection for $y < 12$, especially just above zero.

If the errors are approximately normally distributed and the model is correct, the plot of residuals against fitted values should be without any features, apart from those

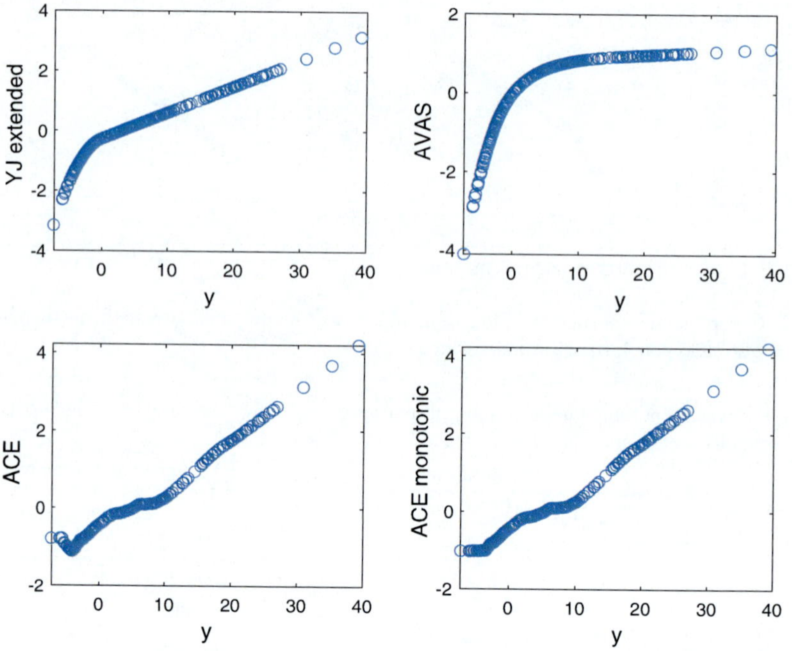

Fig. 3 Contaminated data: transformed responses against untransformed responses. Top row, EYJ and AVAS. Bottom row, ACE, constrained and unconstrained

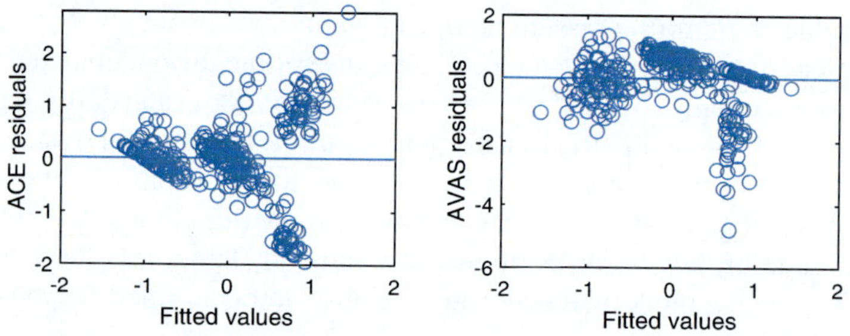

Fig. 4 Contaminated data; residuals against fitted values. Left-hand panel, constrained ACE. Right-hand panel AVAS (note the scale of these residuals)

from the distribution of fitted values. The left-hand panel of Fig. 4 shows such a plot of residuals from constrained ACE. The plot is wedge shaped, with a sharp lower diagonal bound. The other panel, for AVAS, also has some structure, in this case, a cloud of large negative residuals for fitted values around 0.5; the nonparametric transformations indicate faults in the model or data.

We now look at the transformation of the cleaned data after it has been subjected to the extended Yeo–Johnson transformation to check whether the properties can be improved by a further nonparametric transformation. Values of R^2 for such transformations are in the right-hand column of Table 1. The value for EYJ is 0.783.

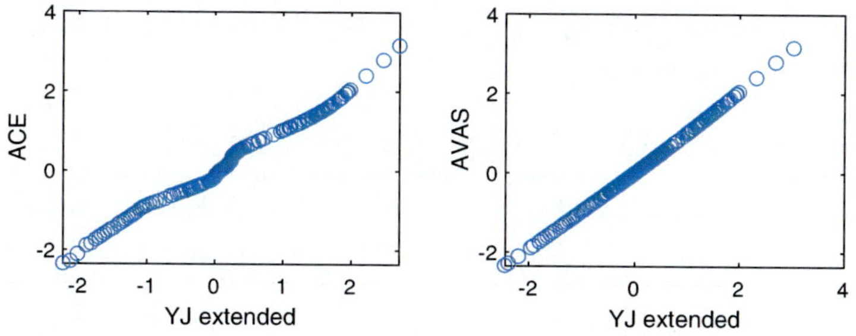

Fig. 5 Nonparametric transformations of cleaned transformed data against EYJ. Left-hand panel, constrained ACE, right-hand panel, AVAS

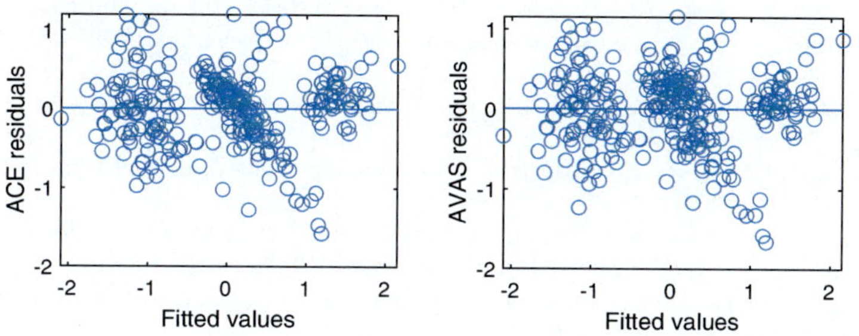

Fig. 6 Residuals against fitted values from nonparametric transformations of cleaned transformed data. Left-hand panel, constrained ACE, right-hand panel, AVAS

AVAS is slightly less than this at 0.778, whereas constrained ACE is 0.805, with the unconstrained version giving a value of 0.806.

The left-hand panel of Fig. 5 shows the plot of transformed y from constrained ACE against the values from EYJ. Some of the points of inflection shown in Fig. 3 remain and correspond to original values of y that were just positive. The indication is that the two-parameter EYJ transformation with one transition point can be improved by using a distinct transformation for the observations just above zero, leading to a slight increase in R^2. The right-hand panel for AVAS shows a virtually straight line and the transformation is very close to that for EYJ.

The plots in Fig. 6 are of residuals against fitted values for the two transformations featured in Fig. 5; both indicate the presence of three groups of funds, which are also surprisingly well transformed by the two-parameter EYJ procedure. Although the two plots are similar, some of the details of the central group are different, which is where the two transformations diverge. The QQ plots for the nonparametric transformations are close to that for EYJ shown in Fig. 2.

The results of this section indicate that the nonparametric transformations do not provide a robust procedure. But they can provide insight when used to check a suggested parametric transformation. For the EYJ it is possible that the two transformation regions may not separate at zero, but at some value to be determined. A second aspect is whether two regions of transformation are enough. It may be that for some data structures the flexibility of the nonparametric transformation will lead to improved data modelling.

Acknowledgements This research benefits from the HPC (High Performance Computing) facility of the University of Parma. We acknowledge financial support from the University of Parma project "Statistics for fraud detection, with applications to trade data and financial statements" and from the Department of Statistics, London School of Economics.

References

1. Atkinson, A.C., Riani, M., Cerioli, A.: The Forward Search: theory and data analysis (with discussion). J. Korean Stat. Soc. **39**, 117–134 (2010). https://doi.org/10.1016/j.jkss.2010.02.007
2. Atkinson, A.C., Riani, M., Corbellini, A.: The analysis of transformations for profit-and-loss data. J. R. Stat. Soc. C **69**, 251–275 (2020)
3. Box, G.E.P., Cox, D.R.: An analysis of transformations (with discussion). J. R. Stat. Soc., Ser. B **26**, 211–246 (1964)
4. Breiman, L., Friedman, J.H.: Estimating optimal transformations for multiple regression and transformation (with discussion). J. Am. Stat. Assoc. **80**, 580–619 (1985)
5. Riani, M., Atkinson, A.C., Cerioli, A.: Finding an unknown number of multivariate outliers. J. R. Stat. Soc., Ser. B **71**, 447–466 (2009)
6. Tibshirani, R.: Estimating transformations for regression via additivity and variance stabilization. J. Am. Stat. Assoc. **83**, 394–405 (1988)
7. Yeo, I.-K., Johnson, R.A.: A new family of power transformations to improve normality or symmetry. Biometrika **87**, 954–959 (2000)

On Predicting Principal Components Through Linear Mixed Models

Simona Balzano, Maja Bozic, Laura Marcis, and Renato Salvatore

Abstract This work introduces a Principal Component Analysis of data given by the Best Predictor of a multivariate random vector. The mixed linear model framework offers a comprehensive baseline to get a dimensionality reduction of a variety of random-effects modeled data. Alongside the suitability of using model covariates and specific covariance structures, the method allows the researcher to assess the crucial changes of a set of multivariate vectors from the observed data to the Best Predicted data. The estimation of the parameters is achieved using the extension to the multivariate case of the distribution-free Variance Least Squares method. An application to some Well-being Italian indicators shows the changeover from longitudinal data to the subject-specific best prediction by a random-effects multivariate Analysis of Variance model.

Keywords Best prediction · Linear mixed model · Variance least squares estimation · Random-effects MANOVA model

1 Introduction

Principal Component Analysis (PCA) is one of the best established methods for dimension reduction. Principal Components (PCs) lead to a better assessment of the available information, by summarizing and visualizing data, and at the same time, minimizing the loss of information [6, 7].

S. Balzano (✉) · M. Bozic · L. Marcis · R. Salvatore
Università di Cassino e del Lazio Meridionale, Cassino, Italy
e-mail: s.balzano@unicas.it

M. Bozic
e-mail: m.bozic@unicas.it

L. Marcis
e-mail: laura.marcis@unicas.it

R. Salvatore
e-mail: rsalvatore@unicas.it

S. Balzano et al. (eds.), *Statistical Learning and Modeling in Data Analysis*, Studies in Classification, Data Analysis, and Knowledge Organization,
https://doi.org/10.1007/978-3-030-69944-4_3

Given a p-variate centered random vector $\mathbf{y}_i$ $(i = 1, \ldots, n)$ and an $n \times p$ matrix of observed data $\mathbf{Y}$ from $\mathbf{y}$, the PCA of $\mathbf{y}$ can be obtained by a Singular Value Decomposition (SVD) of $\mathbf{Y}$ into the matrix product $\mathbf{Y} = \mathbf{P}\mathbf{L}_s\mathbf{Q}' + \mathbf{N} = \mathbf{C}^s\mathbf{Q}' + \mathbf{N}$, where: (i) $\mathbf{P}$ is the s-reduced rank orthogonal matrix of the first s eigenvectors (the left singular vectors) of the symmetric matrix $\mathbf{YY}'$ $(r = 1, \ldots, s, \ldots, p, \quad s \ll p)$, (ii) $\mathbf{L}_s$ is the diagonal matrix of the first s singular values, and (iii) $\mathbf{Q}$ is the s-reduced rank matrix of the eigenvectors (the right singular vectors) of the symmetric covariance matrix $\mathbf{S}_y = \frac{1}{n}\mathbf{Y}'\mathbf{Y}$. The $n \times s$ matrix $\mathbf{C}^s = \mathbf{P}\mathbf{L}_s$ gives the first s principal components, and the $n \times p$ matrix $\mathbf{N}$ reports the cross-product minimum norm matrix of residuals. Given the s-dimensional subspace representation of the observed data, we have $\|\mathbf{N}'\mathbf{N}\|^2 = tr(\mathbf{N}'\mathbf{N}) = \min$ (here tr is the trace of a square matrix).

For decades, PCA has undergone many generalizations and adjustments to the needs of specific research goals. One of them brings into play the role of prediction by the linear statistical models. Bair et al. [1] provided a *supervised* PCA to address the high dimensional issue that arises when the number of predictors, p, far exceeds the number of observations, n-seeking linear combinations with both high variance and significant correlation with the outcome.

Tipping and Bishop [13] had already introduced the notion of prediction for the PCs. They called Probabilistic PCA (probPCA) the model behind the PCA, in which parameters are estimated by means of the Expectation-Maximization algorithm. The "noisy" PC model (nPC), proposed by Ulfarsson and Solo (see [13, 14] for details) has a quite similar formulation respect to the probPC model, providing—in a similar way—the nPC prediction once the model estimates have been given [2, 10].

Unlike the fixed effects PCs, as the traditional linear regression PCA model assumes, the probPC (or nPC) are random variables. This condition suggests, on the one hand, the adoption of the Bayesian approach to handle the estimates for the probPC linear model and, on the other hand, to predict PCs under its meaning within the random linear models theory [9].

The Bayesian approach to the estimation requires an expectation of some model parameters that are random, conditionally to the observed data. Given normality of the error $\boldsymbol{\varepsilon} \sim N(0, \sigma^2\mathbf{I})$, for a linear model $\tau = \mathbf{B}\lambda + \boldsymbol{\varepsilon}$—in case of the vector λ random—the likelihood is based on the conditional distribution $\lambda|\tau \sim N[E(\lambda|\tau), var(\lambda|\tau)]$. Moreover, it is known [8, 9, 11] that $E(\lambda|\tau) = \widetilde{\lambda}$ is the Best Prediction (BP) estimate, with $var(\widetilde{\lambda} - \lambda) = E_\tau[var(\lambda|\tau)]$. This is somewhat different from the standard linear regression model, where the prediction is given by $E(\tau|\lambda)$. Therefore, given a Linear Mixed Model (LMM) for τ, with $E(\tau|\lambda)) = \lambda$, the model parameters become realizations of random variables. The BP of a linear combination of the LMM fixed and random effects (i.e., linear in τ, with $E[E(\tau|\lambda)] = 0$) gives the Best Linear Unbiased Prediction (BLUP) estimates [3, 8, 11].

LMM's are particularly suitable for modeling with covariates (fixed and random) and for specifying model covariance structures [3]. They allow researchers to take into account special data, such as hierarchical, time-dependent, correlated, covariance patterned models. Thus, given the BP estimates of the nPC λ, $\widetilde{\lambda} = E(\lambda|\tau)$, the vector $\widetilde{\tau} = \mathbf{B}\widetilde{\lambda}$ represents the best prediction of the p-variate vector (in the way of the BP).

In general, it is convenient to employ the LMM's to assess how the most relevant parameters affect the linear model assumed for $\mathbf{y}_i$: we acknowledge the difficulty of including in the probPC model some of the typical LMM parameters. For this reason, this work proposes to reverse the BP estimation typical of the probPC model, in the sense that the data from the p-vector may produce itself the BP estimates $\widetilde{\mathbf{y}}_i$ by a multivariate BLUP. Afterwards, ordinary PCs can be obtained by the matrix of the n realizations $\widetilde{\mathbf{y}}_i$. Using the predictive variance of $(\mathbf{y}_i - \widetilde{\mathbf{y}}_i)$, we can configure a double set of analyses analogous to the Redundancy Analysis [12, 15], the last based on the eigenvalue-eigenvector decomposition of the multivariate regression model predictions and errors. Therefore, we have a *constrained* analysis, based on the eigenvalue-eigenvector decomposition of $cov(\widetilde{\mathbf{y}}_i)$, and an *unconstrained* analysis of the Best Prediction model error covariance, $cov(\mathbf{y}_i - \widetilde{\mathbf{y}}_i)$.

The main advantage with respect to Redundancy Analysis is that the novel method may works also without model covariates. This is because the largest part of the multidimensional variability is due to the covariance of the same random effects among the components of the multivariate data vectors. We call this analysis a *predictive* PCA (predPCA), because the PCs are given by the BP data vectors of the subjects.

The proposed procedure would be particularly worthwhile with typically correlated observations, like repeated measures surveys, clustered, longitudinal, and spatially correlated multivariate data. Although the PCA operates only as a final step, this type of analysis can be valuable when the reduction of dimensionality aims to be investigated on data predicted by the sample, rather than the PCA of the sample data by themselves. Usually, the BLUP estimation of the p-variate random effects request iterative procedures in case of likelihood-based methods: the larger is the number of the model parameters, the more computationally expensive is to obtain the estimates to the normal variate covariance components of the LMM model.

Given that the general BLUP estimator has the same form of the BP under normality [8, 11], it is proposed to estimate the model covariance parameters, defining a distribution-free estimator of the BLUP. We introduce a multivariate extension of the Variance Least Squares (VLS) estimation method [4] for the variance components. Because of the specific aspects related to the multivariate case, this method changes from non-iterative to iterative, depending on alternating the minimization procedure from knowing, in turn, one of the two covariance matrices involved in the linear model. For this reason, we obtain an iterative version of the VLS: the Iterative Variance Least Squares (IVLS) method.

When the linear model for $\mathbf{y}_i$ is a population model without fixed covariates, the predPCA is equivalent to a PCA of the n realizations of the p-vector, $\widetilde{\mathbf{y}}_i$. Thus, the linear mixed model is a Multivariate Analysis of Variance (MANOVA) with variance components.

The paper is organized as follows: the first part is dedicated to the predPCA method, together with some explanations about the IVLS estimation. Then, an application of the predPCA method to some Well-being Italian indicators is presented. Two Appendices report some backgrounds and the proof of the Lemma given in the paper.

2 Predictive Principal Components Analysis

Given a p-variate random vector $\mathbf{y}_{ij}$, $i = 1, ..., m$, $j = 1, ..., k$, consider the case when $\mathbf{y}$ is partitioned in m subjects, each of them with k individuals (balanced design). If $\mu' = (\mu_1, ..., \mu_p)$ is the vector of the p means, a random-effects MANOVA model is given by

$$\mathbf{y}_{ij} - \mu = \mathbf{a}_i + \mathbf{e}_{ij}, \quad (1)$$

where $\mathbf{a}_i \overset{ind}{\sim} N_p(0, \Sigma_a)$ is the p-variate random effect and $\mathbf{e}_{ij} \overset{ind}{\sim} N_p(0, \Sigma_e)$ is the model error. Given $n = m \times k$ data from $\mathbf{y}$, we write the model (1) in the LMM standard matrix form $\mathbf{Y} = \mathbf{XB} + \mathbf{ZA} + \mathbf{E}$, where $\mathbf{Y}$ is the $n \times p$ matrix of data from $\mathbf{y}$, $\mathbf{X}$ is a $n \times l$ matrix of explanatory variables, $\mathbf{B}$ the $l \times p$ matrix of the l fixed effects, $\mathbf{Z}$ the $n \times m$ design matrix of random effects, $\mathbf{A}$ is the $m \times p$ matrix of random effects, $\mathbf{E}$ the $n \times p$ matrix of errors.

For the random-effects MANOVA model (1), we have that $\mathbf{X}$ is a column of ones (i.e., $l = 1$), and $\mathbf{B}$ the row vector $\overline{\mu}'$ of sample means:

$$\mathbf{Y} - \mathbf{1}_{n\times 1}\overline{\mu}'_{1\times p} = (\mathbf{I}_m \otimes \mathbf{1}_k) \times (\mathbf{a}_1 ..., \mathbf{a}_p)_{m\times p} + \mathbf{E}, \quad (2)$$

where $\otimes$ is the Kronecker product, $\mathbf{Z} = (\mathbf{I}_m \otimes \mathbf{1}_k)$, $\mathbf{A} = (a_1, ..., a_r, ..., a_p)$. Furthermore, the data $\mathbf{Y}$ and the error matrices have the structure

$$\mathbf{Y}_{mk\times p} = (\mathbf{y}_{11}, \mathbf{y}_{12}, ..., \mathbf{y}_{1k}, ..., \mathbf{y}_{m1}, \mathbf{y}_{m2}, ..., \mathbf{y}_{mk})'$$
$$\mathbf{E}_{mk\times p} = (\mathbf{e}_{11}, \mathbf{e}_{12}, ..., \mathbf{e}_{1k}, ..., \mathbf{e}_{m1}, \mathbf{e}_{m2}, ..., \mathbf{e}_{mk})'.$$

By centering the data $\mathbf{Y}$, with $\mathbf{Y} - \mathbf{1}_{n\times 1}\overline{\mu}'_{1\times p} = \mathbf{Y}^*$, and remembering that $E(\overline{\mu}) = \mu$, the p-vector population model (1) becomes $\mathbf{y}^*_{ij} = \mathbf{a}_i + \mathbf{e}_{ij}$. The BP estimation of the p-vector $\mathbf{a}_i$ in the LMM is given by [3, 8, 11]

$$\widetilde{\mathbf{a}}_i = E(\mathbf{a}_i|\mathbf{y}^*_i) = cov(\mathbf{a}_r, \mathbf{y}^*_i)[var(\mathbf{y}^*_i)]^{-1}[\mathbf{y}^*_i - E(\mathbf{y}^*_i)] \quad (3)$$

Reducing the LMM to the random-effects MANOVA model, we have by the Eq. (2): $E(\mathbf{y}_i) = \mathbf{B}'\mathbf{x}_i = \mu$. It is well-known [8] that the variance of the LMM model is $cov[vec(\mathbf{Y})] = \mathbf{V} = \mathbf{D} + \mathbf{U}$, with $\mathbf{D} = \mathbf{Z} \times cov[vec(\mathbf{A})] \times \mathbf{Z}'$ and $\mathbf{U} = cov[vec(\mathbf{E})]$. The variance matrix $\mathbf{V}$ allows to define a variety of typical linear models, by setting the parameters vector $\theta = (\theta_1, ..., \theta_q)$ inside the components $\mathbf{D}$ and $\mathbf{U}$. The estimation of these parameters is done by standard methods (e.g., Maximum Likelihood, Restricted Maximum Likelihood, Moment Estimator). Given the parameters estimate $\widehat{\theta}$, and then the variance $\widehat{\mathbf{V}} = \mathbf{V}(\widehat{\theta})$, the fixed effects estimate is given by the General Least Squares estimate $\widehat{\mathbf{B}} = \widehat{\mathbf{B}}_{GLS} = (\mathbf{X}'\mathbf{V}^{-1}\mathbf{X})^{-1}X'\mathbf{V}^{-1}\mathbf{Y}^*$. The random effects (3) estimate $\widetilde{\mathbf{A}} = (\widetilde{\mathbf{a}}_1 ..., \widetilde{\mathbf{a}}_r, ..., \widetilde{\mathbf{a}}_p)$, $\widetilde{\mathbf{a}}_r = \text{col}(\widetilde{\mathbf{a}}_{ri})$, $r = 1, ..., p$, completes the so-called Empirical BLUP (EBLUP) $\widetilde{\mathbf{Y}}^* = \mathbf{X}\widehat{\mathbf{B}} + \mathbf{Z}\widetilde{\mathbf{A}}$. We assume for the model (2) the more simple structure, with a single random effect by the i-th subject. Furthermore, an equicorrelation between these random effects is employed.

Some further computational details for the specification of the model (2) are given in Appendix 1.

We introduce an iterative multivariate variance least squares estimation (IVLS) for the estimation of the vector of parameters θ. The objective function to minimize is $VLS = trace(\Xi - \mathbf{U} - \mathbf{D})^2$, with $\Xi_{|mkp \times mkp}$ the empirical model covariance matrix. The algorithm is based on alternating least squares in a two-step iterative optimization process. At every iteration, the IVLS procedure first fixes **U** and the solves for **D**, and then it fixes **D** and solves for **U**. Since the LS solution is unique, at each step the VLS function can either decrease or stay unchanged but never increase. Alternating between the two steps iteratively guarantees convergence only to a local minimum, because it ultimately depends on the initial values for **U**. Being Ξ the matrix of the multivariate *OLS* cross-products of residuals, the VLS iterations are given by the following steps: (a) starting from the separate subject (group)-specific empirical covariance matrices $\mathbf{U}_{ri}$, first minimize VLS to obtain the estimate of the random-effects covariance **D**, then (b), given the matrix $\widehat{\mathbf{B}}_{GLS}\%$, minimize VLS, setting the same error covariance matrix among the subjects, and (c), iterate (a) and (b), until convergence to the minimum. The number of iterations may vary, depending on the choice of the specific model variance structure for the random effects and error covariance matrices.

Applications of the predPCA may be related to different types of available data, and then may accommodate a variety of patterned covariance matrices. Further, groups can be dependent or independent, even in space, time, and space-time correlated data.

The IVLS estimator at each step is unbiased, as discussed in the following Lemma:

Lemma (Unbiasedness of the IVLS estimator) *Under the balanced p -variate variance components MANOVA model $\mathbf{Y}^* = \mathbf{Z}\mathbf{A} + \mathbf{E}$, with $\mathbf{Z}$ the design matrix of random effects, $\mathbf{E}$ the matrix of errors, and covariance matrix $\mathbf{D} + \mathbf{U}$, $\mathbf{D} = (\mathbf{I} \otimes \mathbf{Z})cov[vec(\mathbf{A})](\mathbf{I} \otimes \mathbf{Z}')$, $\mathbf{U} = cov[vec(\mathbf{E})]$, and known matrix $\mathbf{U}$, for the IVLS estimator of the parameters θ in $\mathbf{D}$ we have $E[\mathbf{D} = \mathbf{D}(\widehat{\theta}_{IVLS})] = \mathbf{D}(\theta)$.*

The proof is given in Appendix 2.

Finally, a SVD of the matrix $\widetilde{\mathbf{Y}}$ from the p-dimensional $\widetilde{\mathbf{y}}$ vector is obtained, in order to give a PC decomposition of the subject data involved by the linear model. The predPC are generated by the eigenvalue-eigenvector decomposition of the covariance matrix of the predicted data, i.e., $(\widetilde{\mathbf{Y}} - \mathbf{X}\mathbf{B}(\widehat{\theta}))'(\widetilde{\mathbf{Y}} - \mathbf{X}\mathbf{B}(\widehat{\theta}))$.

3 An Application to Some Well-Being Indicators

The introduced predPCA is applied here for the analysis of some Equitable and Sustainable Well-being indicators (BES), annually provided by the Italian Statistical Institute [16].

The discussed IVLS estimation procedure is adopted.

Table 1 IVLS fixed effects estimates of the random-effect MANOVA model (centered data)

	$vec(\widehat{\beta}_{OLS})$	$vec(\widehat{\beta}_{GLS})$
Education and training	–1.26E-15	–0.008118
Job satisfaction	–2.44E-16	0.0082529
GDP	5.468E-16	0.0024191
Lack of safety	1.062E-16	–0.01079
Research and innovation	–9.35E-16	–0.00471

Table 2 Iterative variance least squares estimates of the random-effects MANOVA model

	IVLS estimates
$\widehat{\sigma}_a^2$	0.374155
$\widehat{\rho}_a$	–0.147169
$\widehat{\sigma}_e^2$	0.242975
$\widehat{\rho}_e$	0.328184
$\widehat{\rho}_t$	0.266346

According to recent law reforms, these indicators should contribute to define the economic policies which largely affect some fundamental dimensions of the quality of life. In this case study, we present an application of predPCA to 5 of the 12 BES indicators available in the years 2013–2016, collected at the level of NUTS2 (Nomenclature of Territorial Units for Statistics). We use the random-effect MANOVA model, where the random multivariate vector **Y** includes the repeated observations of all the Italian regions in the 4 time instants (**X**). We do not consider model covariates, allowing predictors to be derived only by the covariance structure. We assume equicorrelation both of the multivariate random effects and of the residual covariance (see Appendix 1 for details). The random-effects MANOVA model is then given by a balanced design, with an AR(1) error structure.

The fixed effects estimates, obtained through both the OLS and GLS estimators, are provided in Table 1. We have that the GLS estimates outperform the OLS estimates in terms of coefficient's interpretability. The GLS estimate of the variable "Lack of Safety" highlights the greater change in value respect to the OLS mean estimate. This means that this indicator plays the most important role in highlighting the adjustment provided by the model prediction with respect to the observed data. Furthermore, this implies that the Lack of Safety will be the most influential indicator in terms of shifting the statistical units (i.e., the administrative Regions) from their observed position in the factorial plane.

Table 2 shows the IVLS estimation results of the mixed MANOVA model parameters, reporting the estimated variance and correlation among indicators (σ_a, ρ_a) and regression errors (σ_e, ρ_e), in the Σ_a and Σ_e matrices, respectively. We find a negative covariance between the BES indicators, together with a positive covariance between the regression errors among indicators. Finally, the time autocorrelation between

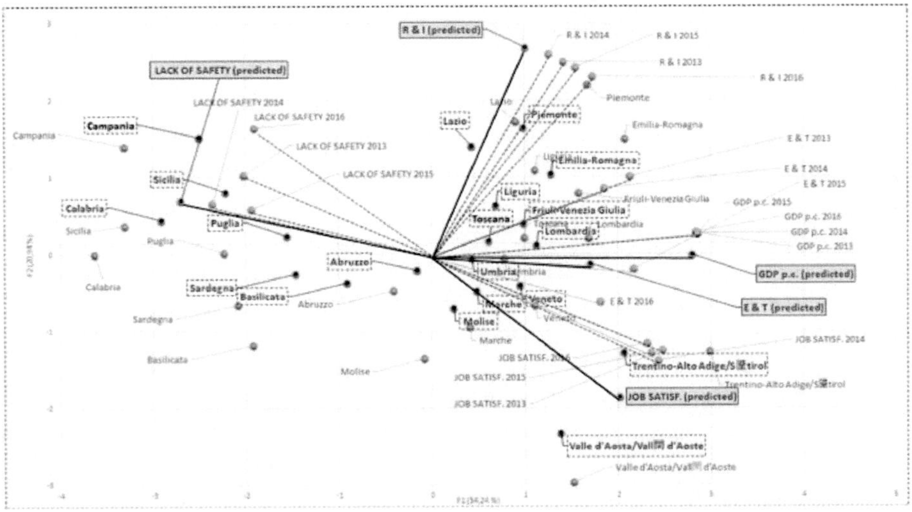

Fig. 1 Multiple Factor Analysis (MFA), observed factor loadings and scores per year (dashed lines); predicted loadings and scores (plain lines) in the space of the MFA

units is estimated as slightly positive, independently from the nature of the BES indicator.

Finally, in order to visualize simultaneously the first factorial axes of the four years on a common factorial plane, for both observed and predicted variables, we performed a Multiple Factor Analysis (MFA) on a matrix obtained by juxtaposing the BES indicators with their IVLS prediction. Figure 1 shows the MFA biplot, where *observed* factor loadings and scores for each year (dashed lines) and *predicted* loadings and scores (plain lines) for each indicator are jointly represented with the *observed* and *predicted* (in rectangles) regions.

On this plan, it is possible to see how the axes change over years (among groups), and at the same time, to foresee how they *could* change in a new situation (in this example on a new year), comparing the position of the observed variable with their IVLS prediction.

Looking at the biplot, the horizontal axis clearly represents the well-being, being positively correlated with the variables GDP, Education and training (E&T), Job satisfaction and Investment in research and development (R&I), and having the variable Lack of Safety always a high negative coordinate. As expected, the Southern Italian regions are concentrated on the left side of the plane.

What is interesting to see is that most of the Southern regions, e.g., Puglia, Campania, Sicily, show a general improvement in terms of *predicted* values along this axis: the coordinates generally move towards the origin, foreseeing a decrease in the Lack of Safety, (i.e., an increase in their Well-being).

4 Conclusions and Perspectives

This paper introduces PCA of a multivariate predictor to perform an exploratory survey of sample data. The predPCA provides a new tool for interpreting a factorial plan, by enriching the factorial solution with the projection of the trends included in the observations. Given a multivariate vector with independent groups, and a random-effects population model, the predPCA relies on the assumption that the linear model itself is able to predict accurately specific subjects or group representatives, even in time and spatial dependent data. The use of the PCA is given afterward when the model has provided data predictions. Substantially, predPCA is a model-based PCA where the data are supplied by the model best predictors.

The advantage in using the predPCA, with respect to the PC-based models, is given by accommodating more easily a variety of structured data by the linear model itself. After using a linear mixed model, the PredPCA explores predicted data that originates in part from the regressive process and in part from the observed ones to understand the contribution of the observed to predictions.

We note that this approach is able to work out simultaneously the issues related to the use of model covariates and specific patterned covariance matrices. The impact of choosing the model structure is easily recognizable when we investigate changes in the factor data description. The reduction of dimensionality of the Best Prediction of a variety of linear models, some of them designed for grouped and correlated data, represents an important issue.

A forthcoming careful consideration will be made against Common Principal Components [5], as a comparative study in terms of a simultaneous representation of different data submatrices. Future studies can accommodate spatial and spatio-temporal data, bringing out the predictive ability of the general linear mixed models, by pivoting on specific covariance structures of the data.

Appendix 1

To accommodate a variety of random effects and error covariance matrices, it is appropriate to refer to the general LMM, as the generalization of the MANOVA variance components model given by Eq. (1):

$$\mathbf{Y} = \widetilde{\mathbf{X}}\mathbf{B} + \widetilde{\mathbf{Z}}\mathbf{A} + \mathbf{E}.$$

We use the vector operator $vec(\mathbf{S})$, that converts the matrix $\mathbf{S}$ in a column vector. Then we have $\mathbf{y} = vec(\mathbf{Y}) = \mathbf{X}\beta + \mathbf{Za} + \mathbf{e}$, $\mathbf{y}_{mkp\times 1} = vec(\mathbf{Y}_{mk\times p})$, $\widetilde{\mathbf{X}}_{mk\times 1} = \mathbf{1}'_p \otimes \mathbf{1}_{mk}$, $\mathbf{B}_{1\times p} = (\beta_{01}, ..., \beta_{0p})$, $\mathbf{X}_{mkp\times p} = \mathbf{I}_p \otimes \mathbf{X} = \mathbf{I}_p \otimes \mathbf{1}_{mk}$, $\beta = vec(\mathbf{B}_{1\times p})$, $\widetilde{\mathbf{Z}}_{mk\times pm} = \mathbf{1}'_p \otimes \mathbf{Z}_r$, $\mathbf{Z}_i = \mathbf{1}_k$, $\mathbf{Z}_r = \mathbf{I}_m \otimes \mathbf{Z}_i = \mathbf{I}_m \otimes \mathbf{1}_k$, $\mathbf{Z}_{p(mk\times m)} = diag(\mathbf{Z}_1, ..., \mathbf{Z}_p)$, $\mathbf{A}_{mp\times p} = diag(\mathbf{a}_1, ..., \mathbf{a}_p)$, $\mathbf{a}_r = \mathrm{col}(\mathbf{a}_{r1}, ..., \mathbf{a}_{rm})$, $\mathbf{a}_{pm\times 1} = \mathrm{col}(\mathrm{col}(\mathbf{a}_{r1}, ..., \mathbf{a}_{rm}))$, and $\mathbf{E}_{mk\times p} = (\mathbf{e}_1, ..., \mathbf{e}_p)$, $\mathbf{e} = \mathrm{col}(\mathbf{e}_1, ..., \mathbf{e}_p) = \mathrm{col}(\mathrm{col}(\mathrm{col}(\mathbf{e}_{rm1}, ..., \mathbf{e}_{rmk})))$.

The BLUP for the j-th group (subject) and r-th response variable is given by $\widetilde{\mathbf{a}}_{ri} = E(\mathbf{a}_{ri}|\mathbf{y}_{ri}) = cov(\mathbf{a}_{ri}, \mathbf{y}_{ri})[var(\mathbf{y}_{ri})]^{-1}[\mathbf{y}_{ri} - E(\mathbf{y}_{ri})]$, with $\mathbf{U}_{ri}$ the covarince matrix of the residual errors for the i-th group and the r-th variable ($r = 1, ..., p$). The fixed effects estimates are given by the matrix $\widehat{\mathbf{B}}_{GLS} = (\mathbf{X}'\mathbf{V}^{-1}\mathbf{X})^{-1}\mathbf{X}'\mathbf{V}^{-1}\mathbf{Y}$, where $\mathbf{V}$ is the model covariance. In the case the variance components MANOVA model (1), if $\mathbf{G}$ is the $p \times p$ covariance matrix of random effects, with $\mathbf{D} = \mathbf{ZGZ}'_{mkp\times mkp} = \mathbf{G} \otimes \mathbf{Z}_r\mathbf{Z}'_r$, $\mathbf{U}_{ri} = \sigma^2_{ri}\mathbf{I}_k$, $\mathbf{U}_r = diag(\mathbf{U}_{r1}, ..., \mathbf{U}_{rm})$, $\mathbf{U}_{mkp\times mkp} = diag(\mathbf{U}_1, ..., \mathbf{U}_p)$, and the model covariance matrix $\mathbf{V}_{mkp\times mkp} = cov(vec\mathbf{Y}) = cov(\mathbf{y}) = \mathbf{ZGZ}' + \mathbf{U} = \mathbf{D} + \mathbf{U}$, we get a "constrained" *PCA* by the predictors, as the SVD of the estimates $\mathbf{Y} - \mathbf{1}\widehat{\mu}'_{GLS} = (\mathbf{I}_m \otimes \mathbf{1}_k) \times (\widetilde{\mathbf{a}}_1, ..., \widetilde{\mathbf{a}}_p)$. Further, an "unconstrained" analysis by the scores of the model conditional residuals $\mathbf{Y} - \widetilde{\mathbf{Y}} = \mathbf{Y} - \mathbf{1}\widehat{\mu}'_{GLS} - (\mathbf{I}_m \otimes \mathbf{1}_k) \times (\widetilde{\mathbf{a}}_1, ..., \widetilde{\mathbf{a}}_p)$ is done. To get the BLUP estimates $\widetilde{\mathbf{a}}_{ri}$, we must know the parameters of the MANOVA model inside the covariance matrix $\mathbf{D} = \mathbf{Z} \times cov(vec(\mathbf{A})) \times \mathbf{Z}'_{mkp\times mkp}$, that is equal to:

$$\mathbf{D} = \Sigma_a \otimes (\mathbf{I}_m \otimes \mathbf{1}_k)(\mathbf{I}_m \otimes \mathbf{1}'_k) = \Sigma_a \otimes (\mathbf{I}_m \otimes \mathbf{1}_k\mathbf{1}'_k).$$

Then: $vec(\mathbf{Y}) = (\mathbf{I}_p \otimes 1_{mt})vec(\mathbf{B}) + (\mathbf{I}_p \otimes Z)vec(\mathbf{A}) + vec(\mathbf{E})$; $\mathbf{y}^* = vec(\mathbf{Y})$, $\mathbf{X}^* = (\mathbf{I}_p \otimes \mathbf{X}) = (\mathbf{I}_p \otimes \mathbf{1}_{mt})$, $\beta^* = vec(\mathbf{B})$, $\mathbf{Z}^*\mathbf{a}^* = (\mathbf{I}_p \otimes \mathbf{Z})vec(\mathbf{A})$.

Further, given the IVLS estimates $\widehat{\theta}$, we have $cov[(\mathbf{y}^*(\widehat{\theta}))] = (\mathbf{I}_p \otimes \mathbf{I}_m \otimes \mathbf{1}_k)(\Sigma_a(\widehat{\theta}) \otimes \mathbf{I}_m)(\mathbf{I}_p \otimes \mathbf{I}_m \otimes \mathbf{1}'_k) + cov(vec(\mathbf{E})) = \Sigma_a(\widehat{\theta}) \otimes (\mathbf{I}_m \otimes \mathbf{1}_k\mathbf{1}'_k) + (\Sigma_e(\widehat{\theta}) \otimes \mathbf{I}_n) \otimes \Omega(\widehat{\theta})$. Finally, after the iterative *VLS* estimation, the predictor is given by $\widetilde{\mathbf{y}}^*(\widehat{\theta}) = \mathbf{X}^*\widehat{\beta}^*_{GLS} + \mathbf{Z}^*\widetilde{\mathbf{a}}^* = \Gamma\mathbf{y}^*(\widehat{\theta}) + (\mathbf{I} - \Gamma)\mathbf{X}^*\widehat{\beta}^*_{GLS}$, $\Gamma = (\Sigma_a(\widehat{\theta}) \otimes \mathbf{ZZ}')cov[(\mathbf{y}^*(\widehat{\theta}))]^{-1}$. Note that the matrix Γ specifies both the contribution of the regression model and the observed data to the prediction.

We assume equicorrelation both of the multivariate random effects and the residual covariance, together with the *AR(1)* structure of the error:

$$\Sigma_a = \sigma_a^2 \times \begin{bmatrix} 1 & \rho_a & \cdots & \rho_a \\ \rho_a & 1 & \cdots & \rho_a \\ \vdots & \cdots & \ddots & \vdots \\ \rho_a & \rho_a & \cdots & 1 \end{bmatrix}_{5\times 5} \qquad \Sigma_e = \sigma_e^2 \times \begin{bmatrix} 1 & \rho_e & \cdots & \rho_e \\ \rho_e & 1 & \cdots & \rho_e \\ \vdots & \cdots & \ddots & \vdots \\ \rho_e & \rho_e & \cdots & 1 \end{bmatrix}_{5\times 5}$$

$$\Omega = \frac{1}{1-\rho_t^2}\begin{pmatrix} 1 & \rho_t & \rho_t^2 & \rho_t^3 \\ \rho_t & 1 & \rho_t & \rho_t^2 \\ \rho_t^2 & \rho_t & 1 & \rho_t \\ \rho_t^3 & \rho_t^2 & \rho_t & 1 \end{pmatrix}_{4\times 4}$$

Appendix 2

Lemma (Unbiasedness of the IVLS estimator) *Under the balanced* p *-variate variance components MANOVA model* $\mathbf{Y}^* = \mathbf{ZA} + \mathbf{E}$, *with* $\mathbf{Z}$ *the design matrix* *of random effects,* $\mathbf{E}$ *the matrix of errors, and covariance matrix* $\mathbf{D}+\mathbf{U}$, $\mathbf{D}=(\mathbf{I}\otimes\mathbf{Z})cov[vec(\mathbf{A})](\mathbf{I}\otimes\mathbf{Z}')$, *with known matrix* $\mathbf{U}=cov[vec(\mathbf{E})]$, *for the* IVLS *estimator of the vector of parameters* θ *in* $\mathbf{D}$ *we have* $E[\mathbf{D}=\mathbf{D}(\widehat{\theta}_{IVLS})]=\mathbf{D}(\theta)$.

Proof With m groups ($i=1,...,m$), each of k individuals ($j=1,...,k$), for the multivariate mixed model we have the vector representation $\mathbf{y}=\mathbf{X}^*\beta+\mathbf{Z}^*\mathbf{a}+\mathbf{e}$, with $\mathbf{y}=vec(\mathbf{Y})$, $\mathbf{X}^*=(\mathbf{I}\otimes\mathbf{X})$, $\beta=vec(\mathbf{B})$, $\mathbf{Z}^*=(\mathbf{I}\otimes\mathbf{Z})$, $\mathbf{a}=vec(\mathbf{A})$, $\mathbf{e}=vec(\mathbf{E})$, and $\eta=\mathbf{Z}^*\mathbf{a}+\mathbf{e}$, $\widehat{\mathbf{B}}=\widehat{\mathbf{B}}_{OLS}$. Defining $\widehat{\boldsymbol{\varepsilon}}=\mathbf{y}-\mathbf{X}^*\widehat{\beta}=\mathbf{X}^*\beta+\eta-\mathbf{X}^*\widehat{\beta}=\eta-\mathbf{X}^*(\widehat{\beta}-\beta)$, by standard results on multivariate regression we write $\widehat{\beta}-\beta=\left\{\mathbf{I}\otimes(\mathbf{X}'\mathbf{X})^{-1}\mathbf{X}'\right\}\mathbf{y}-\beta=\mathbf{C}\times(\mathbf{X}^*\beta+\eta)-\beta$. Thus: $\mathbf{X}^*(\widehat{\beta}-\beta)=\mathbf{X}^*\mathbf{C}\mathbf{X}^*\beta+\mathbf{X}^*\mathbf{C}\eta-\mathbf{X}^*\beta$, and noticing that $\mathbf{C}\mathbf{X}^*=\left\{\mathbf{I}\otimes(\mathbf{X}'\mathbf{X})^{-1}\mathbf{X}'\right\}\mathbf{X}^*=\left\{\mathbf{I}\otimes(\mathbf{X}'\mathbf{X})^{-1}\mathbf{X}'\right\}(\mathbf{I}\otimes\mathbf{X})=\mathbf{I}\otimes(\mathbf{X}'\mathbf{X})^{-1}\mathbf{X}'\mathbf{X}=\mathbf{I}$, we get: $\mathbf{X}^*(\widehat{\beta}-\beta)=\mathbf{X}^*\beta+\mathbf{X}^*\mathbf{C}\eta-\mathbf{X}^*\beta=\mathbf{X}^*\mathbf{C}\eta$, and $\widehat{\varepsilon}=\mathbf{y}-\mathbf{X}^*\widehat{\beta}=\eta-\mathbf{X}^*\mathbf{C}\eta$.

Setting for the MANOVA model $\mathbf{Y}^*=\mathbf{Y}-\mathbf{X}\mathbf{B}$, $\mathbf{X}=\mathbf{1}_{mk\times 1}$, $\mathbf{B}=\mu'_{1\times p}$, to stack matrices by ordering subjects (groups), assume $\mathbf{y}^{**}=vec(\mathbf{Y}^{*\prime})=(\mathbf{Z}\otimes\mathbf{I})vec(\mathbf{A})+vec(\mathbf{E}')=\mathbf{Z}^*\mathbf{a}+\mathbf{e}=\eta$, with $\mathbf{Z}^*$ the design matrix of the multivariate random effects. Given $\widehat{\boldsymbol{\varepsilon}}=vec(\mathbf{Y}'-\widehat{\mathbf{B}}'\mathbf{X}')=\widehat{\mathbf{y}}^{**}$, $\widehat{\mathbf{B}}=\widehat{\mathbf{B}}_{OLS}=\mu'$, the *VLS* estimator finds the minimum of $VLS(\theta)=tr(\mathbf{T}^2)=tr\left\{\widehat{\boldsymbol{\varepsilon}}\widehat{\varepsilon}'-cov(vec(\eta)\right\}^2=\Sigma\mathbf{T}^2_{ij}$. Now denoting $cov(\mathbf{a})=\mathbf{G}=\mathbf{G}(\theta)$, $\mathbf{g}^*=vec(\mathbf{G})$, $\mathbf{u}^*=vec(\mathbf{U})$, and differentiating the *VLS* function with respect to $\mathbf{G}$, we have the following derivatives:

$$\frac{\partial}{\partial\mathbf{G}}VLS(\theta)=\mathbf{Z}^{*\prime}\widehat{\varepsilon}\widehat{\varepsilon}'\mathbf{Z}^*-\mathbf{Z}^{*\prime}\mathbf{Z}^*\mathbf{G}\mathbf{Z}'\mathbf{Z}-\mathbf{Z}^{*\prime}\mathbf{U}\mathbf{Z}^*=0$$
$$(\mathbf{Z}^{*\prime}\mathbf{Z}^*\otimes\mathbf{Z}^{*\prime}\mathbf{Z}^*)\mathbf{g}^*+(\mathbf{Z}^{*\prime}\otimes\mathbf{Z}^{*\prime})\mathbf{u}^*=(\mathbf{Z}^{*\prime}\widehat{\boldsymbol{\varepsilon}})\otimes(\mathbf{Z}^{*\prime}\widehat{\varepsilon}).$$

Then: $\widehat{\mathbf{g}}^*=\mathbf{g}^*(\widehat{\theta})=(\mathbf{Z}^{*\prime}\mathbf{Z}^*\otimes\mathbf{Z}^{*\prime}\mathbf{Z}^*)^{-1}\left\{(\mathbf{Z}^{*\prime}\widehat{\boldsymbol{\varepsilon}})\otimes(\mathbf{Z}^{*\prime}\widehat{\boldsymbol{\varepsilon}})-(\mathbf{Z}^{*\prime}\otimes\mathbf{Z}^{*\prime})\mathbf{u}^*\right\}$.

Remembering that $(\mathbf{Z}^{*\prime}\widehat{\boldsymbol{\varepsilon}})\otimes(\mathbf{Z}^{*\prime}\widehat{\varepsilon})=(\mathbf{Z}^{*\prime}\eta)\otimes(\mathbf{Z}^{*\prime}\eta)$, $cov(\mathbf{a},\mathbf{e})=0$, and taking the expectation of $\eta\otimes\eta$:

$$\begin{aligned}E(\eta\otimes\eta)&=E(vec(\eta\eta'))=E(vec\left\{(\mathbf{Z}^*\mathbf{a}+\mathbf{e})(\mathbf{Z}^*\mathbf{a}+\mathbf{e})'\right\})\\&=E\left\{(\mathbf{Z}^*\otimes\mathbf{Z}^*)vec(\mathbf{a}\mathbf{a}')+(\mathbf{e}\otimes\mathbf{Z}^*)\mathbf{a}+(\mathbf{Z}^*\otimes\mathbf{e})\mathbf{a}+vec(\mathbf{e}\mathbf{e}')\right\}\\&=(\mathbf{Z}^*\otimes\mathbf{Z}^*)\mathbf{g}^*+0+0+\mathbf{u}^*.\end{aligned}$$

Since $(\mathbf{Z}^{*\prime}\eta)\otimes(\mathbf{Z}^{*\prime}\eta)=(\mathbf{Z}^{*\prime}\otimes\mathbf{Z}^{*\prime})(\eta\otimes\eta)$, the expectation become:

$$\begin{aligned}&E\left\{(\mathbf{Z}^{*\prime}\widehat{\boldsymbol{\varepsilon}})\otimes(\mathbf{Z}^{*\prime}\widehat{\boldsymbol{\varepsilon}})\right\}\\&=E\left\{(\mathbf{Z}^{*\prime}\eta)\otimes(\mathbf{Z}^{*\prime}\eta)\right\}=(\mathbf{Z}^{*\prime}\otimes\mathbf{Z}^{*\prime})E(\eta\otimes\eta)\\&=(\mathbf{Z}^{*\prime}\otimes\mathbf{Z}^{*\prime})(\mathbf{Z}^*\otimes\mathbf{Z}^*)\mathbf{g}^*+(\mathbf{Z}^{*\prime}\otimes\mathbf{Z}^{*\prime})\mathbf{u}^*\\&=(\mathbf{Z}^{*\prime}\mathbf{Z}^*\otimes\mathbf{Z}^{*\prime}\mathbf{Z}^*)\mathbf{g}^*+vec(\mathbf{Z}^{*\prime}\mathbf{U}\mathbf{Z}^*).\end{aligned}$$

Hence: $E[\mathbf{g}^*(\widehat{\theta}_{IVLS})]=(\mathbf{Z}^{*\prime}\mathbf{Z}^*\otimes\mathbf{Z}^{*\prime}\mathbf{Z}^*)^{-1}\left\{E\left[(\mathbf{Z}^{*\prime}\widehat{\varepsilon})\otimes(\mathbf{Z}^{*\prime}\widehat{\boldsymbol{\varepsilon}})\right]-(\mathbf{Z}^{*\prime}\otimes\mathbf{Z}^{*\prime})\mathbf{u}^*\right\}=(\mathbf{Z}^{*\prime}\mathbf{Z}^*\otimes\mathbf{Z}^{*\prime}\mathbf{Z}^*)^{-1}\left\{(\mathbf{Z}^{*\prime}\mathbf{Z}^*\otimes\mathbf{Z}^{*\prime}\mathbf{Z}^*)\mathbf{g}^*+vec(\mathbf{Z}^{*\prime}\mathbf{U}\mathbf{Z}^*)-(\mathbf{Z}^{*\prime}\otimes\mathbf{Z}^{*\prime})\mathbf{u}^*\right\}=\mathbf{g}^*(\theta)$.

References

1. Bair, E., Hastie, T., Paul, D., Tibshirani, R.: Prediction by supervised principal components. J. Am. Stat. Assoc. **101**(473), 119–137 (2006)
2. Bartholomew, D.J.: Latent Variable Models and Factor Analysis. Griffin, London (1987)
3. Demidenko, E.: Mixed Models: Theory and Applications. Wiley, New York (2004)
4. Davidian, M., Carroll, R.J.: Variance function estimation. J. Am. Stat. Assoc. **82**, 1079–1091 (1987)
5. Flury, B.N.: Common Principal Components and Related Multivariate Models. Wiley, Inc., New York (1988)
6. Jackson, J.: A User Guide to Principal Components. Wiley, New York (1991)
7. Jolliffe, I.T.: Principal Components Analysis. Springer, New York (2002)
8. McCulloch, C.E., Searle, S.R.: Generalized Linear and Mixed Models. Wiley, New York (2001)
9. Robinson, G.K.: That BLUP is a good thing: the estimation of random effects. Stat. Sci. **6**(1), 15–32 (1991)
10. Schneeweiss, H.: Factors and principal components in the near spherical case. Multivar. Behav. Res. **32**(4), 375–401 (1997)
11. Searle, S.R.: The matrix handling of BLUE and BLUP in the mixed linear model. Linear Algebra Its Appl. **264**, 291–311 (1997)
12. Takane, Y., Jung, S.: Regularized partial and/or constrained redundancy analysis. Psychometrika **73**(4) (2008)
13. Tipping, M.E., Bishop C.M.: Probabilistic principal component analysis. J. R. Stat. Soc., Ser. B (Stat. Methodol.) **61**(3), 611–622 (1999)
14. Ulfarsson, M.O., Solo, V.: Sparse variable PCA using geodesic steepest descent. IEEE Trans. Signal Process. **56**(12), 5823–5832 (2008)
15. van den Wollenberg, A.L.: Redundancy analysis an alternative for canonical correlation analysis. Psychometrika **42**, 207–219 (1977)
16. https://www.istat.it/en/well-being-and-sustainability

Robust Model-Based Learning to Discover New Wheat Varieties and Discriminate Adulterated Kernels in X-Ray Images

Andrea Cappozzo, Francesca Greselin, and Thomas Brendan Murphy

Abstract In semi-supervised classification, class memberships are learnt from a trustworthy set of units. Despite careful data collection, some labels in the learning set could be unreliable (label noise). Further, a proportion of observations might depart from the main structure of the data (outliers) and new groups may appear in the test set, which were not encountered earlier in the training phase (unobserved classes). Therefore, we present here a robust and adaptive version of the Discriminant Analysis rule, capable of handling situations in which one or more of the aforementioned problems occur. The proposed approach is successfully employed in performing anomaly and novelty detection on geometric features recorded from X-ray photograms of grain kernels from different varieties.

Keywords Impartial trimming · Label noise · Model-based classification · Novelty detection · Anomaly detection · Robust estimation

1 Introduction and Motivation

Thanks to scientific advances, sophisticated techniques like X-ray, scanning microscopy and laser technology are increasingly employed for automatic imaging collection. Unfortunately, among the many observations obtained via measurement and record-

A. Cappozzo (✉) · F. Greselin
Department of Statistics and Quantitative Methods, University of Milano-Bicocca, Milano, Italy
e-mail: a.cappozzo@campus.unimib.it

F. Greselin
e-mail: francesca.greselin@unimib.it

T. B. Murphy
School of Mathematics & Statistics and Insight Research Centre, University College Dublin, Dublin, Ireland
e-mail: brendan.murphy@ucd.ie

S. Balzano et al. (eds.), *Statistical Learning and Modeling in Data Analysis*, Studies in Classification, Data Analysis, and Knowledge Organization,
https://doi.org/10.1007/978-3-030-69944-4_4

ing, some unreliable units may appear: the percentage of encoding errors in real-world databases, all fields taken together, is estimated to be approximately five percent [8]. Therefore, there is strong interest in developing methodologies that perform reliable inference even when standard assumptions are not met, as it happens when dealing with complex contaminated datasets. In discriminant analysis, for example, it is assumed that a set of outlier-free and correctly labeled units are available for each and every group within the population of interest. Nevertheless, this may not hold true, for instance, in image classification, where data quality is influenced by the number of pixels in each sample and by the variability associated with the labeling task [16]. Moreover, as more and more units are acquired, previously unseen structures may emerge.

Motivated by a dataset recording geometric parameters of grains, detected using a soft X-ray technique, we propose a new method for anomaly and novelty detection. Specifically, we introduce a robust model-based approach for adaptive classification: novelties are assumed to arise from a mixture of multivariate normal densities, while no distributional assumption is a priori set for the anomalies. Robustness, based on trimming the least likely observations, copes with training units whose class memberships are unreliable (label noise) and with specimens that are far away from the main data structure (outliers). On the other hand, groups not previously encountered within the labeled units (unobserved classes) are easily added in the form of new mixture components by adaptive learning.

The rest of the paper is organized as follows. In Sect. 2 the notation is introduced and the main concepts about the model and its inferential aspects are presented. In Sect. 3 we apply our methodology to discriminate different varieties of wheat kernels, under adulteration and sample selection bias. Section 4 summarizes the novel contributions and concludes the manuscript.

2 RAEDDA Model

Let us consider a classification framework with $\{(\mathbf{x}_1, \mathbf{l}_1), \ldots, (\mathbf{x}_N, \mathbf{l}_N)\}$ identifying the training set: $\mathbf{x}_n$ is a p-variate outcome and $\mathbf{l}_n$ its associated class label, such that $l_{ng} = 1$ if observation n belongs to group g and 0 otherwise, $g = 1, \ldots, G$. Correspondingly, let $\{(\mathbf{y}_1, \mathbf{z}_1), \ldots, (\mathbf{y}_M, \mathbf{z}_M)\}$ be the test set, where it is assumed, differently from the standard framework, that the unknown classes $\mathbf{z}_m$ have dimension $E \geq G$. That is, there may be a number H of "hidden" classes in the test, not previously observed within the labeled units, such that $E = G + H$, with $H \geq 0$. Both $\mathbf{x}_n$, $n = 1, \ldots, N$, and $\mathbf{y}_m$, $m = 1, \ldots, M$, are assumed to be independent realizations of a continuous random vector $\mathscr{X}$ taking values in $\mathbb{R}^p$; while $\mathbf{l}_n$ and $\mathbf{z}_m$ are considered to be realizations of a discrete random vector $\mathscr{C}$ taking values in $\{1, \ldots, E\}$. Notice that we implicitly suppose here that an unknown sample selection bias mechanism prevents the learning units to arise from classes $G + 1, \ldots, E$. Assuming a Gaussian mixture distribution for $\mathscr{X}$, the *observed data likelihood* reads:

$$L(\boldsymbol{\tau}, \boldsymbol{\mu}, \boldsymbol{\Sigma}|\mathbf{X}, \mathbf{Y}, \mathbf{l}) = \prod_{n=1}^{N}\prod_{g=1}^{G}\left[\tau_g \phi(\mathbf{x}_n; \boldsymbol{\mu}_g, \boldsymbol{\Sigma}_g)\right]^{l_{ng}} \prod_{m=1}^{M}\left[\sum_{g=1}^{E}\tau_g \phi(\mathbf{y}_m; \boldsymbol{\mu}_g, \boldsymbol{\Sigma}_g)\right] \tag{1}$$

where τ_g is the prior probability of observing class g, such that $\sum_{g=1}^{E}\tau_g = 1$, and $\phi(\cdot; \boldsymbol{\mu}_g, \boldsymbol{\Sigma}_g)$ represents the multivariate Gaussian density with mean vector $\boldsymbol{\mu}_g$ and covariance matrix $\boldsymbol{\Sigma}_g$. Notice that the first term in (1) accounts for the complete observations $(\mathbf{x}_n, \mathbf{l}_n)$; whereas in the second term only the marginal density of $\mathbf{y}_m$ contributes to the product, since its associated label $\mathbf{z}_m$ is unknown. Equation (1) defines the likelihood of an Adaptive Mixture Discriminant Analysis (AMDA) model, introduced in [2]. By means of impartial trimming [10], patterned covariance matrices [1, 5] and constrained parameter estimation [11], we extend the original AMDA method developing a flexible classifier, denoted Robust and Adaptive Eigenvalue Decomposition Discriminant Analysis (RAEDDA), which performs reliable supervised classification when dealing with label noise, outliers and unobserved classes. RAEDDA parameters are obtained by maximizing the *trimmed observed data log-likelihood:*

$$\begin{aligned}\ell_{trim}(\boldsymbol{\tau}, \boldsymbol{\mu}, \boldsymbol{\Sigma}|\mathbf{X}, \mathbf{Y}, \mathbf{l}) = &\sum_{n=1}^{N}\zeta(\mathbf{x}_n)\sum_{g=1}^{G}\mathbf{l}_{ng}\log\left(\tau_g\phi(\mathbf{x}_n; \boldsymbol{\mu}_g, \boldsymbol{\Sigma}_g)\right) + \\ &+\sum_{m=1}^{M}\varphi(\mathbf{y}_m)\log\left(\sum_{g=1}^{E}\tau_g\phi(\mathbf{y}_m; \boldsymbol{\mu}_g, \boldsymbol{\Sigma}_g)\right)\end{aligned} \tag{2}$$

where $\zeta(\cdot)$ and $\varphi(\cdot)$ are indicator functions that determine whether each observation contributes or not to the trimmed likelihood. The trimming levels α_l and α_u are pre-specified such that only $\sum_{n=1}^{N}\zeta(\mathbf{x}_n) = \lceil N(1-\alpha_l)\rceil$ and $\sum_{m=1}^{M}\varphi(\mathbf{y}_m) = \lceil M(1-\alpha_u)\rceil$ terms are not null in (2). Notice that the total number E of groups is not established in advance and needs to be estimated: a dedicated penalized likelihood criterion, based on the one introduced in [6], is developed for model selection. Two alternative estimation procedures for maximizing (2) are proposed: the transductive and the inductive learning approaches. Computational details are reported in the next subsections.

2.1 *Transductive Learning*

In the transductive approach, the parameters of both known and hidden classes are concurrently estimated via the joint exploitation of training and test sets. That is, labeled and unlabeled units mutually partake in the learning procedure: the maximization of (2) is carried out via an adaptation of the EM algorithm that includes a Concentration step [14] for enforcing impartial trimming and an eigenvalue-ratio restriction [9] for protecting the final estimates from spurious local maximizers.

In detail, each iteration begins with a C-step, in which the $\lfloor N\alpha_l \rfloor$ and $\lfloor M\alpha_u \rfloor$ least likely units (under the currently estimated model) are tentatively discarded in the training and test sets, respectively. Afterwards, in the E-step the expected value of the unknown label for each untrimmed unit $\mathbf{y}_m$ is computed. Then, an M-step is performed: parameters are updated by determining the set $\{\hat{\tau}_g, \hat{\boldsymbol{\mu}}_g, \hat{\boldsymbol{\Sigma}}_g\}$, $g = 1, \ldots, E$, which maximizes the transductive *trimmed complete data log-likelihood*

$$\ell_{trim_c}(\boldsymbol{\tau}, \boldsymbol{\mu}, \boldsymbol{\Sigma}|\mathbf{X}, \mathbf{Y}, \mathbf{l}, \hat{\mathbf{z}}) = \sum_{n=1}^{N} \zeta(\mathbf{x}_n) \sum_{g=1}^{G} l_{ng} \log\left(\tau_g \phi(\mathbf{x}_n; \boldsymbol{\mu}_g, \boldsymbol{\Sigma}_g)\right) + \\ + \sum_{m=1}^{M} \varphi(\mathbf{y}_m) \sum_{g=1}^{E} \hat{z}_{mg} \log\left(\tau_g \phi(\mathbf{y}_m; \boldsymbol{\mu}_g, \boldsymbol{\Sigma}_g)\right) \tag{3}$$

where the $\hat{z}_{mg}$ have been previously determined in the E-step. Lastly, whenever the estimated covariance matrices do not satisfy the eigenvalue-ratio restriction [11], constrained estimation is enforced.

Once convergence is reached, the final output comprises the set of estimated parameters for the E classes, values for the indicator functions $\zeta(\cdot)$ and $\varphi(\cdot)$ that pinpoint unreliable units, and a posteriori classification for the unlabeled observations via the maximum a posteriori (MAP) estimate [12]. For a more comprehensive description of the algorithm, the interested reader is referred to Sect. 3.2 of [4].

2.2 Inductive Learning

In the inductive approach, parameters are determined in a sequential manner: firstly the training set is employed for robustly estimating the structure of the G known classes (robust learning phase) and, subsequently, the extra classes are sought in the test set keeping the structure learnt in the previous step fixed (robust discovery phase). The first phase consists in the robust fitting of a fully supervised model-based classifier: the REDDA method introduced in [3]. In the robust discovery phase, we search for the $H = E - G$ hidden classes in an unsupervised fashion, by maximizing the likelihood on the test set via an EM algorithm. Each iteration begins with a C-step, in which the $\lfloor M\alpha_u \rfloor$ least likely units are tentatively discarded. Notice that both the current estimates for the parameters of the H hidden classes, as well as the structure of the G known groups (previously determined in the learning phase) concur in the determination of the trimming functions. Then, a standard E-step is computed. Afterwards, an M-step is performed: parameters are updated by determining the set $\{\hat{\tau}_1, \ldots, \hat{\tau}_E, \hat{\boldsymbol{\mu}}_{G+1}, \ldots, \hat{\boldsymbol{\mu}}_E, \hat{\boldsymbol{\Sigma}}_{G+1}, \ldots, \hat{\boldsymbol{\Sigma}}_E\}$ that maximizes the inductive *trimmed complete data log-likelihood*:

$$\ell_{trim_c}(\boldsymbol{\tau}, \boldsymbol{\mu}, \boldsymbol{\Sigma}|\mathbf{Y}, \bar{\boldsymbol{\mu}}, \bar{\boldsymbol{\Sigma}}, \hat{\mathbf{z}}) = \sum_{m=1}^{M} \varphi(\mathbf{y}_m)\Bigg(\sum_{g=1}^{G} \hat{z}_{mg} \log(\tau_g \phi(\mathbf{y}_m; \bar{\boldsymbol{\mu}}_g, \bar{\boldsymbol{\Sigma}}_g)) + \\ + \sum_{h=G+1}^{E} \hat{z}_{mh} \log(\tau_h \phi(\mathbf{y}_m; \boldsymbol{\mu}_h, \boldsymbol{\Sigma}_h))\Bigg) \quad (4)$$

where the $\hat{z}_{mg}$ have been determined in the E-step and the parameters for the G known classes, identified by a bar in the notation, were obtained in the learning phase and are therefore kept fixed. Notice that the entire vector $\boldsymbol{\tau}$ is updated, renormalizing the mixing proportions for the G known classes according to the estimated sizes of the H new groups. Once convergence is reached, the output of the discovery phase comprises the set of estimated parameters for the H new classes, values for the indicator function $\varphi(\cdot)$ that pinpoint unreliable test units, and a posteriori classification for the unlabeled observations via the MAP rule. For a more comprehensive description of the algorithm, the interested reader is referred to Sect. 3.3.2 of [4].

3 Anomaly and Novelty Detection in X-Ray Images of Wheat Kernels

The methodology presented in the previous section is employed to perform adaptive classification and anomaly detection in a dataset comprised of 210 grains belonging to three different varieties of wheat. For every sample (70 units for each variety), seven geometric parameters are recorded from postprocessing X-ray photograms of the kernel [7]. The seeds dataset is publicly available in the University of California, Irvine Machine Learning data repository.

The considered experiment involves the random selection of 98 training units from the first two cultivars, and a test set of 102 samples, including 60 grains from the third variety (data are displayed in Fig. 1). The remaining 10 units from the third group are appended to the training set and their associated labels are altered, as to pretend they come from the first variety. Besides, for 7 randomly chosen training units the `length` variable is manually modified to be three times larger than its original value. The aim of the experiment is, therefore, to determine whether the RAEDDA method is capable of recovering the unobserved class in the test set while coping with both class and attribute noise in the training set. The study is repeated $B = 100$ times: for each recurrence, model results for RAEDDA and for the original AMDA model (denoting by RAEDDAt, AMDAt and RAEDDAi, AMDAi their transductive and inductive versions) and for two popular novelty detection methodologies, namely Classifier Instability (QDA-ND) [17] and Support Vector Machine for novelty detection (SVM-ND) [15] are collected.

In Table 1, we report two metrics for evaluating the correct classification rate and the recovery of the true test partition. The RAEDDA model shows a remarkably good classification accuracy: the unseen class is correctly discovered via both transductive

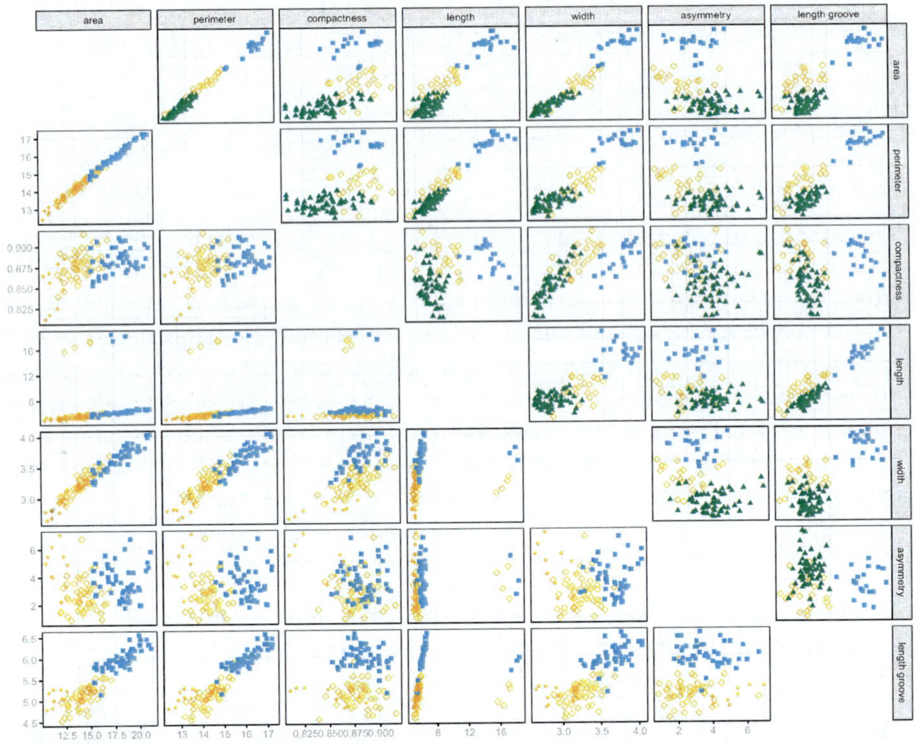

Fig. 1 Learning scenario for the considered experiment, seeds dataset. Plots below the main diagonal represent the training set, in which the first two wheat varieties are displayed with hollow diamonds and solid squares, respectively. Solid diamonds denote the 10 units from the third variety with altered labels. Plots above the main diagonal represent the test set

and inductive inference with the underlying test partition effectively retrieved, as demonstrated by the high average value of the Adjusted Rand Index (ARI) [13]. The AMDA method instead reports a large misclassification error: the outlying units obscures the separation between the first and the third wheat variety.

It is interesting to notice, however, that the test partition is adequately well recovered by AMDA, since its ARI metric presents comparable values to those obtained by our proposal. This intriguing result is explained by looking at the number of estimated components for the two model-based methods, displayed in the barplot of Fig. 2. In trying to mitigate the bias induced by the noise in the learning phase, the non-robust methodology tends to overestimate the true number of hidden classes. On the one hand, this produces a satisfactory clustering in the test set, allowing the model to correctly identify the patterns that were originally contaminated in the training set. On the other hand, estimated parameters for the known classes are highly biased and thus their structure is no longer paired with the (outlier-free) test units: the true varieties are identified as extra classes in the unlabeled set.

Table 1 Average misclassification errors and Adjusted Rand Index for AMDA and RAEDDA classifiers (transductive and inductive inference) and accuracy in separating known and hidden patterns for QDA-ND and SVM-ND on the test set for $B = 100$ runs of the considered experiment, seeds dataset. Standard deviations are reported in parentheses

	RAEDDAt	RAEDDAi	AMDAt	AMDAi	SVM-ND	QDA-ND
Misclassification error	0.082 (0.021)	0.105 (0.073)	0.521 (0.293)	0.43 (0.324)	0.329 (0.185)	0.34 (0.045)
ARI	0.788 (0.052)	0.735 (0.102)	0.674 (0.155)	0.745 (0.105)	–	–

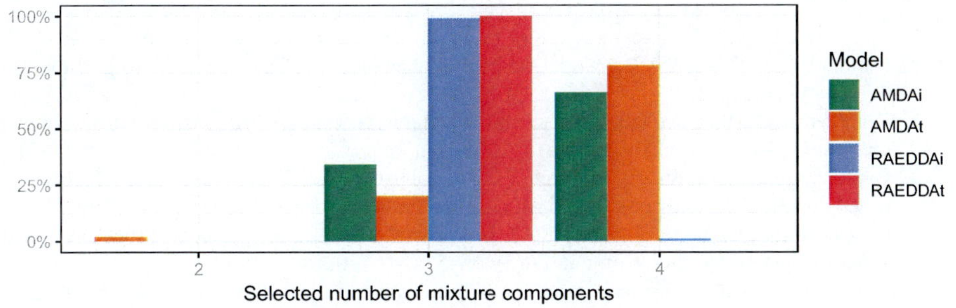

Fig. 2 Percentage of times, out of $B = 100$ runs of the considered experiment, each model-based method identifies the final estimated mixture to have 2, 3 or 4 components. The correct value is 3, as the test set contains the two known classes of wheat, plus the one previously unseen

Low classification accuracy is displayed also by the novelty detection techniques, where the mislabeled units have a severe impact on the correct separation between known and hidden patterns. The same does not happen for our robust proposal, and setting trimming values respectively equal to 0.15 and 0.05 for the training and test sets prevents the noisy units to jeopardize the learning process. The units with inflated length (attribute noise) and 7 out of the 10 wrongly labeled units (class noise) are on average correctly identified to be anomalies, discarding them from the estimation procedure and so yielding higher classification accuracy. Such a result is noteworthy as the separation between the third and first wheat variety is not at all apparent by looking at the pairs plot in Fig. 1.

4 Conclusions

In the present paper, we have introduced a methodology that performs classification in presence of adulteration and sample selection bias. We have employed it in effectively achieving anomaly and novelty detection in X-ray images of grain kernels, where a challenging classification framework, including label noise and outliers, along with one unobserved wheat variety, has been considered.

Further research directions include the extension of the present methodology to high-dimensional classification: a robust and adaptive variable selection procedure, based on theoretical results for Gaussian mixtures, is currently being developed.

Acknowledgements Brendan Murphy's work is supported by Science Foundation Ireland grants SFI/12/RC/2289_P2 and 16/RC/3835. Andrea Cappozzo and Francesca Greselin's work is supported by Milano-Bicocca University Fund for Scientific Research, 2019-ATE-0076.

References

1. Banfield, J.D., Raftery, A.E.: Model-based Gaussian and non-Gaussian clustering. Biometrics **49**(3), 803 (1993)
2. Bouveyron, C.: Adaptive mixture discriminant analysis for supervised learning with unobserved classes. J. Classif. **31**(1), 49–84 (2014)
3. Cappozzo, A., Greselin, F., Murphy, T.B.: A robust approach to model-based classification based on trimming and constraints. Adv. Data Anal. Classif. **14**(2), 327–354 (2020)
4. Cappozzo, A., Greselin, F., Murphy, T.B.: Anomaly and Novelty detection for robust semi-supervised learning. Stat. Comput. **30**(5), 1545–1571 (2020)
5. Celeux, G., Govaert, G.: Gaussian parsimonious clustering models. Pattern Recognit. **28**(5), 781–793 (1995)
6. Cerioli, A., García-Escudero, L.A., Mayo-Iscar, A., Riani, M.: Finding the number of normal groups in model-based clustering via constrained likelihoods. J. Comput. Graph. Stat. **27**(2), 404–416 (2018)
7. Charytanowicz, M., Niewczas, J., Kulczycki, P., Kowalski, P.A., Łukasik, S., Zak, S.: Complete gradient clustering algorithm for features analysis of X-ray images. Adv. Intell. Soft Comput. **69**, 15–24 (2010)
8. Frénay, B., Verleysen, M.: Classification in the presence of label noise: A survey. IEEE Trans. Neural Networks Learn. Syst. **25**(5), 845–869 (2014)
9. Fritz, H., García-Escudero, L.A., Mayo-Iscar, A.: A fast algorithm for robust constrained clustering. Comput. Stat. Data Anal. **61**, 124–136 (2013)
10. Gordaliza, A.: Best approximations to random variables based on trimming procedures. J. Approx. Theory **64**(2), 162–180 (1991)
11. Ingrassia, S.: A likelihood-based constrained algorithm for multivariate normal mixture models. Stat. Methods Appl. **13**(2), 151–166 (2004)
12. McLachlan, G.J.: Discriminant Analysis and Statistical Pattern Recognition, Wiley Series in Probability and Statistics, vol. 544. John Wiley & Sons Inc, Hoboken, NJ, USA (1992)
13. Rand, W.M.: Objective criteria for the evaluation of clustering methods. J. Am. Stat. Assoc. **66**(336), 846 (1971)
14. Rousseeuw, P.J., Driessen, K.V.: A fast algorithm for the minimum covariance determinant estimator. Technometrics **41**(3), 212–223 (1999)
15. Schölkopf, B., Williamson, R., Smola, A., Shawe-Taylor, J., Platt, J.: Support vector method for novelty detection. Adv. Neural Inf. Process. Syst. **12**, 582–588 (2000)
16. Smyth, P., Fayyad, U., Burl, M.: Inferring ground truth from subjective labelling of venus images. Adv. Neural Inf. Process. Syst. **7**, 1085–1092 (1995)
17. Tax, D.M.J., Duin, R.P.W.: Outlier detection using classifier instability. In: A. Amin, D. Dori, P. Pudil, H. Freeman (eds.) Adv. Pattern Recognit., pp. 593–601. Springer Berlin Heidelberg, Berlin, Heidelberg (1998)

A Dynamic Model for Ordinal Time Series: An Application to Consumers' Perceptions of Inflation

Marcella Corduas

Abstract This article discusses an innovative model for time series ordinal data, which develops the well-established CUB model to allow for time-varying parameters. This is a mixture of a Uniform and a Shifted Binomial distribution, characterized by two parameters that can be interpreted as a measure of the ability of the rater to use the available rating scale and the degree of liking/disliking about the item. For illustrative purposes, the method is applied to consumers' perceptions of inflation in Italy.

Keywords Ordinal time series · CUB model · Time-varying model · Qualitative survey data

1 Introduction

Repeated surveys about opinions, perceptions, or attitudes of the interviewees are regularly carried out by national statistical offices. Elementary data are usually not available because individuals are randomly selected each time, and only the aggregate frequency distributions of opinions are published. This is the case of the surveys concerning the qualitative assessment or anticipations on the price level that ISTAT carries out every month.

Measuring public's inflation expectations and perceptions of inflation is of great importance for monetary authorities because both expectations and perceptions are key determinants of actual inflation. For this reason, numerous studies have focused their attention either on quantifying the observed opinion data in order to derive indices of perceived (or expected) inflation or on searching explicative models that could describe data in terms of economic explanatory variables [1, 13]. In this article,

M. Corduas (✉)
Department of Political Sciences, University of Naples Federico II, Naples, Italy
e-mail: marcella.corduas@unina.it

S. Balzano et al. (eds.), *Statistical Learning and Modeling in Data Analysis*, Studies in Classification, Data Analysis, and Knowledge Organization,
https://doi.org/10.1007/978-3-030-69944-4_5

we discuss an innovative model for time series ordinal data, that extends the well-established CUB model to allow for time-varying parameters. The paper is organized as follows. Firstly, we briefly recall the main features of the CUB model. Then, we extend the formulation so that time-varying parameters are allowed. Finally, for illustrative purposes, the method is applied to the time series of consumers' perceptions of inflation in Italy.

2 The Static CUB Model

A class of mixture distributions for ordinal data, denoted as CUB model, has been widely investigated in the past decade, proving its usefulness in numerous empirical studies (see, among others, [9–11]). In particular, ratings are described by a random variable Y characterized by the following probability mass distribution:

$$p(y;\theta) = \pi \binom{m-1}{y-1} (1-\xi)^{y-1}\xi^{m-y} + (1-\pi)\frac{1}{m}, \qquad y = 1, 2, \ldots, m \quad (1)$$

where $\theta = (\pi, \xi)'$, $\xi \in [0, 1]$, $\pi \in (0, 1]$ and $m > 3$. Hence, the parameter space is given by:

$$\Omega(\theta) = \Omega(\pi, \xi) = \{(\pi, \xi) : \quad 0 < \pi \leq 1, 0 \leq \xi \leq 1\}. \quad (2)$$

The weight π determines the contribution of the uniform distribution in the mixture, therefore, $(1-\pi)$ is interpreted as a measure of the ability of the rater to use the available rating scale. This component has been denoted as *uncertainty*. Besides, the parameter ξ characterizes the shifted Binomial distribution and is related to the rater's perception of the item content. For this reason, it has been denoted as *feeling*. Specifically, $(1-\xi)$ denotes the degree of liking/disliking expressed by raters about the item. Assuming that the question is expressed with positive wording and that the lowest score is attached to the worst judgement, when $(1-\xi) > 0.5$ the skewness of the distribution is negative so that the portion of individuals attaching a high rating to the item under evaluation is large. The opposite is verified when $(1-\xi) < 0.5$.

Various developments have been discussed in the literature. In particular, the model has been extended to account for the presence of a 'shelter category', where a respondent refuges himself when he is unwilling to elaborate an accurate judgement [8]. In this case, the random variable Y is described by a GeCUB model such that

$$p(y;\theta) = \delta D^c + (1-\delta)\left[\pi \binom{m-1}{y-1} (1-\xi)^{y-1}\xi^{m-y} + (1-\pi)\frac{1}{m}\right] \quad (3)$$

where D^c is a degenerate distribution at the 'shelter category' c , and $\theta = (\delta, \pi, \xi)'$, with $0 \leq \delta \leq 1$.

In the following section, we will introduce a dynamic version of such a model that can be useful to describe the qualitative assessment of items in repeated surveys.

3 The Dynamic Model for Ordinal Time Series

Let $\{Y_t,\ t = 1, ..., T\}$ be a collection of random variables describing ordinal data observed at different time points. We assume that at time t, the variable Y_t is characterized by the following GeCUB distribution:

$$P(Y_t = y|I_{t-1}) = \delta_t D_t + (1-\delta_t)\left[\pi_t \binom{m-1}{y-1}(1-\xi_t)^{y-1}\xi_t^{m-y} + (1-\pi_t)\frac{1}{m}\right]$$
$$y = 1, 2, ..., m$$

with

$$\pi_t = \frac{1}{1+e^{-\beta_0-\beta_1 z_{t-1}...\beta_p z_{t-p}}};\quad \xi_t = \frac{1}{1+e^{-\gamma_0-\gamma_1 w_{t-1}...\gamma_s w_{t-s}}};$$
$$\delta_t = \frac{1}{1+e^{-\alpha_0-\alpha_1 v_{t-1}...\alpha_k v_{t-k}}};\tag{4}$$

where z_t, w_t and v_t are explanatory variables, I_{t-1}, is the set of information concerning these variables until time $(t-1)$. Moreover, D_t is a degenerate distribution such that: $D_t = 1$ for the shelter category and $D_t = 0$ for the remaining categories. Finally, $\beta = (\beta_0, \beta_1, ..., \beta_p)'$ and $\gamma = (\gamma_0, \gamma_1, ..., \gamma_s)'$, and $\alpha = (\alpha_0, \alpha_1, ..., \alpha_k)'$ are the parameter vectors. Without losing in generality, we concentrate our attention on the case when each GeCUB parameter is affected by one explanatory variable at various lags, but the model can be easily extended so that several explanatory variables are included. Moreover, note that when the shelter effect is not present, the model collapses to the CUB formulation with time-varying parameters.

Let us denote with $[f_{1t}, f_{2t}, ..., f_{mt}]$ the relative frequencies from a random sample of n observations drawn from Y_t, $t = 1, 2, ..., T$. The estimation of the model (4) can be performed by minimizing the sum of the Pearson's chi-square distances between the observed relative frequencies and the GeCUB probabilities:

$$G(\theta) = n\sum_{t=1}^{T}\sum_{y=1}^{m}[f_{yt} - p_{yt}]^2/p_{yt}\tag{5}$$

where the notation has been simplified denoting with $\theta = (\alpha', \beta', \gamma')'$ the vector of $r = k + p + s + 3$ parameters, and $p_{yt} = p_{yt}(\theta) = P(Y_t = y|I_{t-1})$. It is well known that the minimum chi-square method yields estimates that are asymptotically equivalent to maximum likelihood estimates. In particular, they are consistent and asymptotically efficient (see [3] p. 425–6; [7] and references therein). Then, the parameter estimators are asymptotically normal with mean θ and asymptotic variance covariance matrix Q^{-1} with $Q = \{q_{ih}\}$:

$$q_{ih} = -\sum_{t=1}^{T}\sum_{y=1}^{m} n_{yt} p_{yt}(\theta) \frac{\partial^2 \log p_{yt}(\theta)}{\partial\theta_i \partial\theta_h}. \quad (6)$$

being $n_{yt} = nf_{yt}$ the absolute frequencies.

Finally, the goodness of fit of the model is assessed by comparing $\widetilde{G}_{mod} = n^{-1}G(\widehat{\theta})$ with the distance $\widetilde{G}_U = m\sum_{t=1}^{T}\sum_{i=1}^{m}[f_{it} - m^{-1}]^2$. This measures the discrepancy of the observed frequencies from the uniform probabilities, which reflects the situation of pure ignorance about the phenomenon under investigation.

The model (4) can be used for various purposes. Firstly, the dynamic pattern of the estimated parameters helps to detect how the ordinal distributions change over time. In our opinion, this characterization is more informative with respect to the study of the time series of a certain summary statistics (for example, the mean) of the empirical distribution observed at time t. Secondly, the model is useful for predicting the probability distribution of Y_{T+k} using the past realizations (or predictions) of the explanatory variables. Finally, by analogy to the static model, the pattern of the estimated time-varying parameters can be exploited to compare the dynamics of various ordinal time series.

4 A Case Study: Consumer Inflation Perceptions

Consumers' qualitative opinions about the development of inflation are regularly surveyed by ISTAT within the harmonized European programme of business and consumer surveys. Specifically, in Italy, every month a sample of about 2000 consumers are interviewed about their perceptions of past inflation development and their expectations about the future. The first variable, Y_t, is originated from the question (Q5): 'How do you think that consumer prices have developed over the last 12 months? They have: risen a lot; risen moderately; risen slightly; stayed about the same; fallen'. The second one, Z_t refers to the question (Q6): 'By comparison with the past 12 months, how do you expect consumer prices will develop in the next 12 months? They will: increase more rapidly; increase at the same rate; increase at a slower rate; stay about the same; fall'. Only the frequency distribution of the opinion categories is published monthly. In this section, we analyze data ranging from 1994.01 to 2018.1. A preliminary study of this data-set has been presented by [4]. Here, the observed categories have been recoded so that 1 is associated to the category 'fallen/fall', and 5 to the category 'risen a lot/increase more rapidly'. This scale is reversed with respect to that widely used in the economic literature.

The shape of the distributions of the ordinal variable, Y_t, associated to each time point may vary depending on the economic situation, as Fig. 1 shows. For this reason, the perceived change in inflation is usually evaluated by the balance statistic: $B(t) = b = -f_{1t} - 0.5f_{2t} + 0.5f_{4t} + f_{5t}$. This measure is often compared graphically with the actual inflation rate. In this regard, it is worth recalling that the link between inflation perceptions and actual inflation had been quite strong before 2002, but this

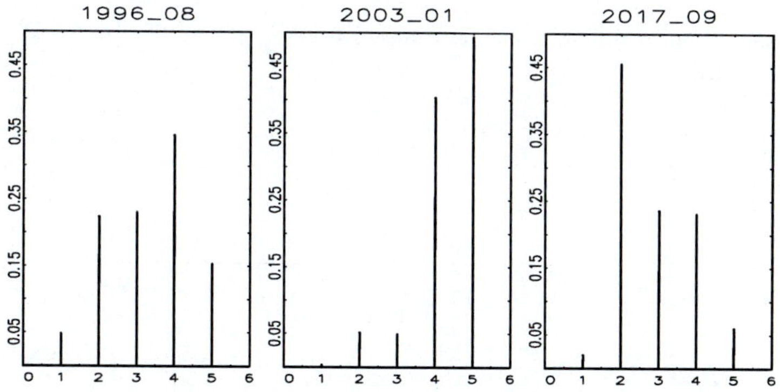

Fig. 1 Examples of observed frequency distribution of Y_t for selected time points

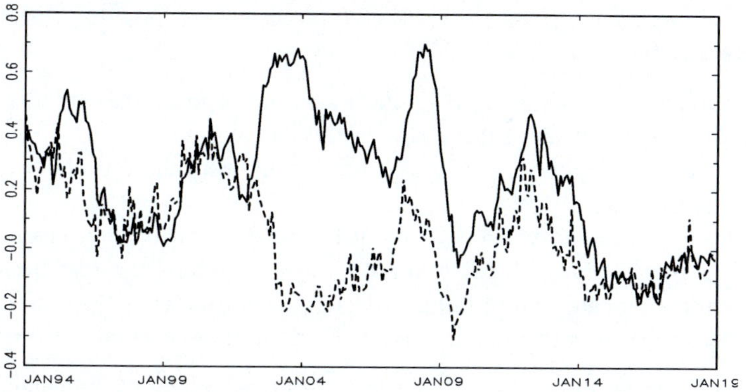

Fig. 2 Balance statistic of perceived (solid line) and expected (dashes) inflation

co-movement disappeared after the Euro cash changeover in 2002 in all EU countries [2]. In Italy, this gap was exceptionally large and persistent, and a similar divergent pattern also affected the balance statistic of perceived and expected inflation [5]. Only towards the end of 2007, perceptions and expectations started again to move together, even if the gap began to reduce only after the 2008 global economic crises (Fig. 2).

We have applied the model (4) to describe the dynamics of ordinal data originated by the question concerning the perception of past price development. A conceptual framework of the process generating consumer's opinions about inflation has been illustrated by [12]. The socio-economic environment, the amplification due to media, and personal attitudes (gender, personal income, level of education) are all important drivers. In addition, the perceptions are strictly related to the expectations. This is not only true from the present to the future, but expectation about the price trend, formed at some previous time, may in some cases bias the perceptions of the current situation [6, 14]. Moving from those considerations, we have specified the dynamics of the GeCUB coefficients as follows:

Table 1 Estimation results (standard errors in parenthesis)

$\hat{\gamma}_0 = 3.213(0.010)$	$\hat{\gamma}_1 = -1.128(0.003)$	Fitting measures
$\hat{\beta}_0 = -2.221(0.129\ )$	$\hat{\beta}_1 = 1.704(0.043)$	$\widetilde{G}_{mod} = 4.04$
$\hat{\alpha}_0 = -1.857(0.006)$	$\hat{\alpha}_1 = -0.724(0.010)$	$\widetilde{G}_U = 157.32$

$$P(Y_t = y|I_{t-1}) = \delta_t D_t + (1-\delta_t)\left[\pi_t \binom{m-1}{y-1}(1-\xi_t)^{y-1}\xi_t^{m-y} + (1-\pi_t)\frac{1}{m}\right], \quad y = 1, 2, ..., 5.$$

$$\xi_t = \frac{1}{1+e^{-\gamma_0-\gamma_1\overline{y}_{t-1}}};\ \pi_t = \frac{1}{1+e^{-\beta_0-\beta_1\overline{z}_{t-1}}};\ \delta_t = \frac{1}{1+e^{-\alpha_0-\alpha_1 v_{t-1}}}; \tag{7}$$

where, for any t:

- the parameter ξ_t depends on $\overline{y}_{t-1}$, the mean of the price past trend perceptions (this is simply the mean of the observed ratings) at time $t-1$;
- the parameter π_t depends on $\overline{z}_{t-1}$, the mean of the expectations about future price level at time $t-1$;
- $D_t = 1$ for the category: 'stayed about the same', and 0 otherwise. The corresponding coefficient δ_t depends on $v_{t-1} = \overline{y}_{t-1} - \overline{z}_{t-1}$, the gap between price trend perceptions and future trend expectations at time $(t-1)$. When this gap is small, the perception that prices stayed about the same becomes stronger.

Table 1 illustrates the estimated coefficients of the model with their standard errors in parenthesis. Computations have been done using the programming system GAUSS (Aptech Systems, Inc.). The global fitting of the model is satisfactory as the remarkable reduction of the discrepancy between the observed and fitted distributions shows. The time plot of $(1-\hat{\xi}_t)$ helps to detect the main characteristics of the distributions of the ordinal variable Y_t (see Fig. 3, panel a). From 1994 to the beginning of 2014, $(1-\hat{\xi}_t) > 0.5$. This implies that most of the estimated ordinal distributions are left skewed because consumers tend to state that prices have increased in the last twelve months. High values of $(1-\hat{\xi}_t)$ are achieved after the Euro cash changeover. Other remarkable fluctuations can be recognized between 2010 and 2013 when various international and national political crises affected financial indicators (such as the increase of the spread between 10-year BTP and German bund) feeding the uncertainty of consumers about the economy. Only at beginning of 2014 the time series collapsed below 0.5 and start to fluctuate around that value.

The pattern of weight of the shelter category $\hat{\delta}_t$ and the weight of the Uniform distribution, $(1-\hat{\pi}_t)(1-\hat{\delta}_t)$ (i.e. uncertainty) are illustrated in Fig. 3, panel b. Both components have a limited role in determining the mixture. However, after the Euro cash changeover, the two components follow an opposite but consistent pattern. As a matter of fact, the role of the 'uncertainty' increases whereas the weight of the refuge and neutral category decreases.

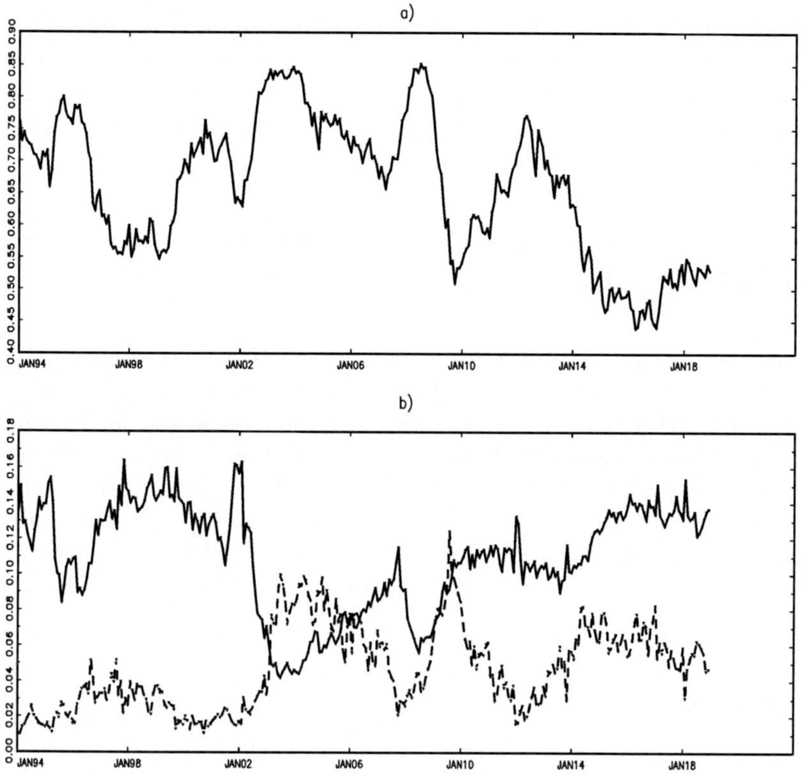

Fig. 3 Time-varying coefficients: **a** $(1-\hat{\xi}_t)$; **b** $\hat{\delta}_t$ (solid line); $(1-\hat{\pi}_t)*(1-\hat{\delta}_t)$ (short dashed)

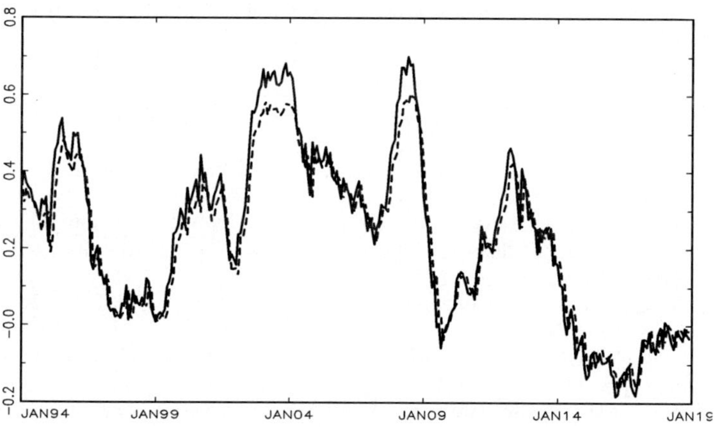

Fig. 4 Balance statistic from the empirical distributions (solid line) and the estimated model (dashed line)

The plot of the observed balance statistic with that implied by the model confirms the goodness of the results (Fig. 4). The two time series are very close in all the considered time interval. In this regard, it is worth to point out that the model is also able to reproduce the large increase that occurred in the time series with the Euro cash changeover.

5 Final Remarks

We have presented a parsimonious model for describing time series of ordinal data that exploits the features of the CUB model. The analysis of the pattern of characterizing parameters helps to summarize the changes in the ordinal distributions along time. Firstly, this synthesis is more informative than using simple summary statistics, such as the average, to describe the dynamics of the phenomenon originating the ordinal data. Secondly, the model provides a useful tool for prediction and control, because the relationships that define the time-varying parameters are specified as a function of explanatory variables for which future scenarios may be elaborated.

References

1. Arioli, R., Bates, C., Dieden, H., Duca, I., Friz, R., Gayer, C., Kenny, G., Meyler, A., Pavlova, I.: EU consumers? quantitative inflation perceptions and expectations: an evaluation. ECB Occasional Paper Series No 186, Frankfurt am Main (D) (2017)
2. Arnold, I.J., Lemmen, J.J.: Inflation expectations and inflation uncertainty in the eurozone: evidence from survey data. Rev. World Econ. **144**, 325–346 (2008)
3. Cramér, H.: Mathematical Methods of Statistics. Princeton University Press, Princeton (1948)
4. Corduas, M., Simone, R., Piccolo, D.: Modelling consumers? qualitative perceptions of inflation. In: Porzio, G., Greselin, F., Balzano, S. (eds), CLADAG 2019 Book of Short Papers, pp. 136–139 (2019)
5. Del Giovane, P., Fabiani, S., Sabbatini, R.: Perceived and measured after the launch of the euro: explaining the gap in Italy. Giorn. Econom. e Ann. Econ. **65**, 155–192 (2006)
6. Greitemeyer, T., Schulz-Hardt, S., Traut-Mattausch, E., Frey, D.: The influence of price trend expectations on price trend perceptions: why the Euro seems to make life more expensive? J. Econ. Psychol. **26**, 541–548 (2005)
7. Harris, R.R., Kanji, G.K.: On the use of minimum chi square estimation. J. R. Stat. Soc. (D) **32**, 379–394 (1983)
8. Iannario, M.: Modelling shelter choices in a class of mixture models for ordinal responses. Stat. Meth. Appl. **21**, 1–22 (2012)
9. Piccolo, D.: Observed information matrix for MUB models. Quaderni di Statistica **8**, 33–78 (2006)
10. Piccolo, D., Simone, R.: The class of cub models: statistical foundations, inferential issues and empirical evidence. Stat. Methods Appl. **28**, 389–435 (2019)
11. Piccolo, D., Simone, R., Iannario, M.: Cumulative and CUB models for rating data: a comparative analysis. Intl. Stat. Rev. **87**, 207–236 (2019)
12. Ranyard, R., Del Missier, F., Bonini, N., Duxbury, D., Summers, B.: Perceptions and expectations of price changes and inflation: a review and conceptual framework. J. Econ. Psychol. **29**, 378–400 (2008)

13. Simmons, P., Weiserbs, D.: Consumer price perceptions and expectations. Oxf. Econ. Pap. **44**, 35–50 (1992)
14. Traut-Mattausch, E., Schulz-Hardt, S., Greitemeyer, T., Frey, D.: Expectancy confirmation in spite of disconfirming evidence: the case of price increases due to the introduction of the Euro. Eur. J. Soc. Psychol. **34**, 739–760 (2004)

Deep Learning to Jointly Analyze Images and Clinical Data for Disease Detection

Federica Crobu and Agostino Di Ciaccio

Abstract In recent years, computer-assisted diagnostic systems increasingly gained interest through the use of deep learning techniques. Surely, the medical field could be one of the best environments in which the power of the AI algorithms can be tangible for everyone. Deep learning models can be useful to help radiologists elaborate fast and even more accurate diagnosis or accelerate the triage systems in hospitals. However, differently from other fields of works, the collaboration and co-work between data scientists and physicians is crucial in order to achieve better performances. With this work, we show how it is possible to classify X-ray images through a multi-input neural network that also considers clinical data. Indeed, the use of clinical information together with the images allowed us to obtain better results than those already present in the literature on the same data.

Keywords Deep learning · Medical deep learning · Convolutional neural networks · X-ray images · Multi-input neural networks

1 Introduction

Recent years have been marked by an exponential growth of interest towards whatever concerns data. Thanks to their great availability and hardware/software breakthroughs, many improvements and progresses have been made in the world of deep learning [5]. The use of deep convolutional neural networks has had a great impact on image recognition techniques. In this context, the evolution of research conducted

F. Crobu (✉) · A. Di Ciaccio
Department of Statistics, Sapienza Università di Roma, Rome, Italy
e-mail: federicacrobu@gmail.com

A. Di Ciaccio
e-mail: agostino.diciaccio@uniroma1.it

S. Balzano et al. (eds.), *Statistical Learning and Modeling in Data Analysis*,
Studies in Classification, Data Analysis, and Knowledge Organization,
https://doi.org/10.1007/978-3-030-69944-4_6

in the past decade is well represented by the continuous progress in the ImageNet dataset [9], until few years ago considered the benchmark for new architectures.

Among the many scopes of the AI, one of the most fascinating and advantageous branches for the application of these models is medicine. However the challenge is more complex: while within the general context of image recognition, the goal is to classify what it is contained in a given image (since the information is completely inside the picture), in the specific case of medical images we also should consider other important information about the patients. In fact, in order to try to emulate the role of an expert radiologist, the model should consider much more information such as demographic and clinical details.

Doctors usually gather and handle all this information and it is equally advantageous to provide them to the predictive model. From the technical point of view, the goal of including more inputs of different nature can be achieved using a multi-input neural network architecture. Using this kind of model we were able to obtain a very accurate classification, as shown in the following sections. Until a few years ago, it was unreasonable to think about a future in which doctors would be helped by computers to recognize diseases and elaborate diagnoses. The impact of these new technologies could represent a drastic improvement in underdeveloped countries, where the availability of doctors can often be problematic and pathologies such as pneumonia are still one of the main causes of death. Moreover, it could also be helpful in wealthy countries, where the number of radiologists is insufficient.

2 State of the Art and Challenges of the Medical Deep Learning

Among the many studies, some stand out for having achieved an accuracy comparable with that of the radiologists. DeepMind and Google Health have successfully trained an algorithm on mammogram images from a large database of 28,953 female patients in the US and the UK, the results were published in the journal Nature [10]. Considering 2 images for each breast, they analysed 115,812 images. In a standard analysis, about 20% of screenings fail to find breast cancer even when it is present and many others are false positive. The AI algorithm decreased both types of error performing better than human radiologists (AUC 0.889 for UK data). To correctly evaluate the results, the real outcomes were derived from the biopsy record and longitudinal follow-up. NYU researchers published a similar study [11] using 229,426 screening mammography exams on 141,473 patients with about 1 million of images. Their network achieved an AUC of 0.895 and, to validate the model, they conducted a reader study with 14 radiologists, each analyzing 720 exams. The model confirmed its goodness showing an accuracy higher than a single experienced radiologist. However, both studies concluded that the AI screenings should be used in tandem with radiologists. In fact, thanks to the combination of experienced doctors and computers, it is possible to obtain the most precise diagnostic results.

The benefits of AI systems in automated triaging of chest radiographs have been explored by Annarumma et al. [2]. This work uses more than four hundred thousand X-rays, jointly with their reports. Firstly, an NLP algorithm extracts the prioritization level from each report, subsequently a DCNN model associates the urgency from the image's analysis. The new prioritization system was tested in a simulation study, which showed a shorter mean delay for critical cases.

To apply these methodologies, an important requirement is the availability of a large and reliable database of images and clinical evaluations. This need clashes with the complexity of the image labeling process, as the definition of diagnosis is always characterized by a certain subjectivity, even if made by expert radiologists.

In general, an optimal solution to the problems faced in the medical area is to include much more information beyond the mere analysis of the images. For example, the correlation of certain pathologies to age or smoking is well known. Other diseases may be characterized by genetic predispositions and many diseases can be related to each other. Thus, the more additional information we have about the patients' clinical history the more we are able to construct a framework useful to improve the predictive model.

This work is based on our previous paper [4], a similar approach, given by Baltruschat [3], is discussed in Sect. 3.

3 Material and Methods

Among the general framework of medical deep learning we decided to focus on the analysis of X-ray images. Probably, the largest public database containing both images and clinical information is ChestX-ray14. From a technical point of view, we had to solve a multi-class and multi-label problem, since the task is the prediction of presence/absence of 14 diseases that can coexist in the same diagnostic image.

We will demonstrate how a multi-input neural network, so called since it is made by two independent nets joined in the end to perform predictions, can fruitfully use the information provided by the images with that coming from the patients' other data.

3.1 The Data

The *ChestX-ray14* [15] database was released in 2017 by the United States National Institutes of Health (NIH) and contains over 112,000 radiographic frontal chest images of 30,805 patients. To exemplify, some of them are displayed in Fig. 1.

Each of them can be healthy or sick, affected by one or more of the following 14 diseases: Atelectasis, Cardiomegaly, Consolidation, Edema, Effusion, Emphysema, Fibrosis, Hernia, Infiltration, Mass, Nodule, Pleural thickening, Pneumonia, Pneumothorax. Furthermore, a "no finding" category represents the images in which none

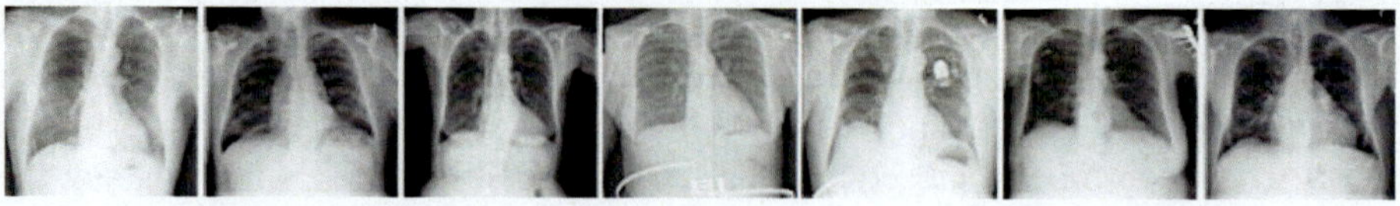

Fig. 1 Some images of the database ChestX-ray14

of the previously mentioned diseases have been detected. "No finding" is the diagnosis in 60361 radiographs, while, for example, the diagnosis of "only" pneumonia is given in 322 cases. Few clinical and demographic information were also available: age, patient gender and follow-up number. In fact, each patient may have performed more than one radiographic examination, the progressive follow-up number indicates the sequence in which the examination was performed.

The labels, corresponding to the pathologies identified in each image, were extracted from radiological reports using natural language processing techniques with an accuracy that is declared by the authors over 90% [15]. Therefore, we cannot fully trust the labelling process and, furthermore, some researchers have raised many doubts about the correctness of the labels. Most of the criticism has been advanced by the radiologist Luke Oakden-Rayner [12] who, after observing the images, stated that many incorrect labels are present in the data and thus he could not say what the algorithm would be really able to understand and learn from such images. In addition, reports are mainly written in order to help other doctors, and the labels extracted by them can be different from the final diagnosis of the physicians.

3.2 Previous Works on the Same Data

This dataset has been already used by many other researchers. Surely, the best-known work was made by a Stanford's team [13]. They proposed an architecture called CheXNet based on the usage of the DCNN architecture called DenseNet-121 [8]. This work represents, at present, the state-of-the-art results in terms of AUC scores. Other important works are Yao's et al. [16] and Wang's et al. [15]. The first one is mainly based on an architecture consisting of a DenseNet as an encoder and on a recurrent neural network as a decoder. Wang tries to apply some of the most famous CNN architectures (excluding DenseNet), achieving the best results with ResNet-50 [7]. However, there are numerous other papers that address this problem on the same or similar data using a deep neural network. For example, [1] proposed to apply a pretrained CNN as a feature extraction from the images and then, in sequence, a classification model. Another interesting work is that of Gündel [6], but these papers did not use additional clinical data. More interesting, from our point of view, is the Baltruschat's [3] work.

The last paper uses a multi-input neural network and includes the analysis of 3 variables: age, sex and view position. The architecture of the model was based on the ResNet-50 model applied to 448x448 px images. Because of this choice, which implies different input sizes than those expected by the model (224x224 px), the authors added as first layer one Max Pooling to reduce the size of the images. The three variables were concatenated and directly linked to the output. In their work, some choices were introduced that impede a direct comparison with other applications in the literature. In particular, they did not use the 'official partition' (the benchmark train/test split proposed by the authors of the database). Although they experimented different architectures based on the ResNet model, the results obtained seem worse than those of the previous papers. They stressed the importance of including clinical data, but they did not consider the patient's medical history, which, to some degree, could be derived from the data. These aspects and the model architecture constitute the main elements of differentiation from our proposal.

3.3 The Model

Inspired by the Stanford's work, we decided to enrich the model by including the few clinical and demographic information available with these images. To reach our goal, we employed two independent networks that are joined at the end in order to share information before making predictions (a schematic drawing can be seen in Fig. 3).

The first and main branch consists of a Convolutional Neural Network, suitable to capture the essence of the X-rays. Among the many possibilities available, we decided to adopt the DenseNet-121 model, which is a CNN with 121 layers. The aspect that characterizes the architecture is the presence of 4 dense blocks, respectively, with 6-12-24-16 layers inside. The blocks are connected by transition blocks each consisting of one convolutional layer and one pooling layer, which have the task of reducing the dimensionality (see Fig. 2).

The potential of this architecture lies in the usage of a deep structure characterized by many "short paths" between the layers that constitute the network itself [8]. This innovative mechanism lets the information pass directly from a layer to all the other ones, in a feed-forward fashion. This model has shown to be very efficient in terms of optimization, achieving top performances on benchmark datasets as ImageNet.

The second and innovative step is the building of the parallel network which processes the non-image characteristics. It considers age, opportunely rescaled using min-max normalization, sex and other 14 new dummy variables using the follow-up information. In fact, we constructed these new variables by recording patient information obtained in the previous pathological history, if present in the data. This branch of the network includes one input layer with 16 neurons and two hidden dense layers with 128 neurons activated by a ReLU function.

Finally, the two networks are concatenated and connected to the output layer consisting of 14 neurons with sigmoid activation function, whose task is to estimate the probability of the presence of each disease in the X-ray image.

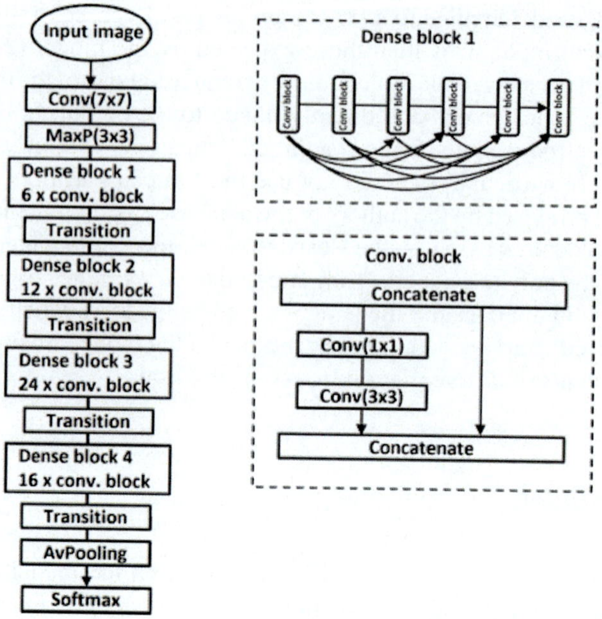

Fig. 2 The DenseNet-121 architecture [8], based on the repetition of two kind of blocks: the *dense block*, able to perform the concatenation of many different convolution filters of different size, and the *transition block*, which performs the compression of the information. In order to make possible the last step, the CNN structure has to be flattened: this is performed using a Global Average Pooling layer

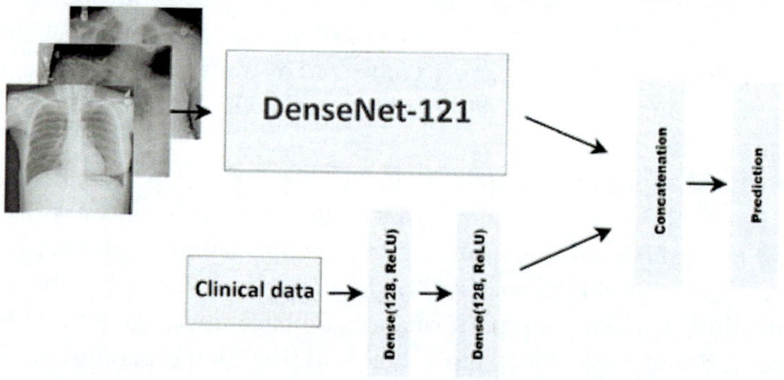

Fig. 3 Multi-input neural network architecture. On the top the DenseNet-121 architecture [8] in which the 'top layers' have been eliminated. The branch at the bottom consists of two hidden dense layers applied to the non-image inputs. The two branches are then concatenated in order to produce predictions

The data was divided using the official benchmark partition proposed by [15], which consists of 80% for the training set and 20% for the test set. To make the tuning of the model we used 20% of the training set as a validation set. The entire network has a complex structure with 123 ‘main’ layers and more than 7 million parameters. We used the pretrained weights of DenseNet-121 (without the top-layers) on Imagenet as initialization of the convolutional neural network, while the second network has been trained from scratch using random weights. In the first epochs, to avoid the corruption of Imagenet’s pre-trained weights, DenseNet’s weights were frozen. To solve this multi-input multi-output problem, we have employed a weighted binary cross-entropy loss function [5] for accounting the high imbalance among the classes. Moreover, a data augmentation has been applied to the images. We tried several alternatives, for example, adding noise or a slight zoom, but a simple horizontal flip of the X-ray resulted to be the best choice. As regard the optimization technique, we have chosen the *Adam* method with a tiny learning rate (0.001 and 0.0001 to fine tune). Adam combines the best properties of the AdaGrad and RMSProp algorithms to provide an optimization algorithm that can handle sparse gradients on noisy problems [5]. To perform the analysis, we used the Tensorflow library and one Nvidia Titan XP 6100 GPU. To train one model it took up to 120 h.

4 The Results

Despite the limited clinical and demographic data available, our approach provides an interesting improvement of the state-of-the-art results, confirming our intuition of the architecture’s power. Following the literature, we have adopted the AUC (Area Under the ROC Curve) index as the main tool to evaluate the quality of the predictions (Fig. 4). Table 1 shows the comparison of the performances of our model with the best results obtained by other researchers in terms of AUC scores.

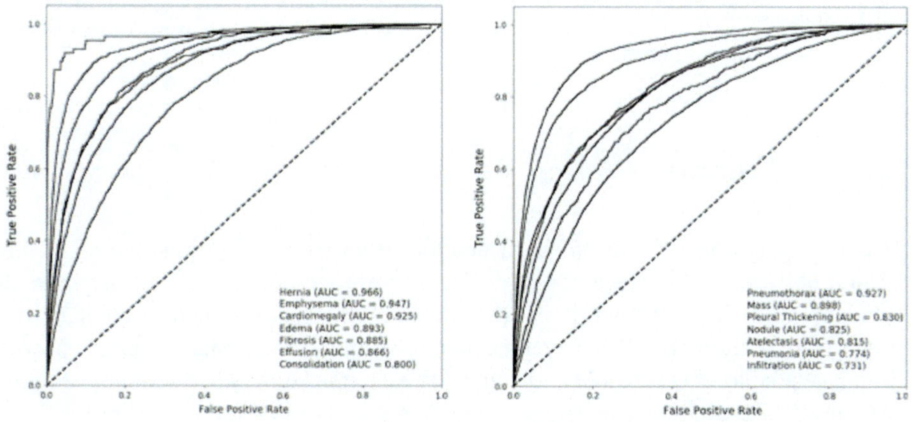

Fig. 4 ROC curves of the 14 diseases on the training (left) and test (right). The diseases’ curves are represented according to the decreasing AUC scores order

Table 1 AUC scores comparison

	Wang et al.	Yao et al.	CheXNet	**Multi-input**
Official split	Yes	No	No	Yes
Atelectasis	0.716	0.772	0.809	**0.815**
Cardiomegaly	0.807	0.904	**0.925**	**0.925**
Effusion	0.784	0.859	0.864	**0.866**
Infiltration	0.609	0.695	**0.735**	0.731
Mass	0.706	0.792	0.868	**0.898**
Nodule	0.671	0.717	0.780	**0.825**
Pneumonia	0.633	0.713	0.768	**0.774**
Pneumothorax	0.806	0.841	0.889	**0.927**
Consolidation	0.708	0.788	0.790	**0.800**
Edema	0.835	0.882	0.888	**0.893**
Emphysema	0.815	0.829	0.937	**0.947**
Fibrosis	0.769	0.767	0.805	**0.885**
Pleural thickening	0.708	0.765	0.806	**0.830**
Hernia	0.767	0.914	0.916	**0.966**
Average	0.738	0.803	0.841	**0.863**

It is evident in Table 1 that the average AUC has been significantly improved by our approach and, for most classes, we have clearly outperformed previous jobs. The scores show great variability: from 0.731 for Infiltrations to 0.966 for Hernia. The reason for these differences can be partly attributed to the imbalance of the data (even if we have applied appropriate weights to the training set), and partly to the differences between the pathologies: some of them are more difficult to identify with the available information.

5 Conclusions

The results of this application confirmed the validity of our approach: a multi-input neural network architecture can significantly improve predictions. Clearly, the idea of combining different heterogeneous sources of information can be applied in many other fields of medicine. Whenever the patient's clinical and/or demographic information is available, it is possible and fruitful to take this approach. Another possibility, which could produce great strides in medical AI, could be the joint real-time work with the radiologist [14]. In this way all the entities involved could enjoy significant advantages: doctors would be helped by the computer while analyzing the images and the algorithm would be trained in real-life situations, making a tangible contribution to its development.

Finally, it would be remarkable to have more public medical data in order to improve the researches, hoping that future studies in this sector will lead to a better quality of life and healthcare all over the world.

References

1. Allaouzi, I., Ahmed, M.B.: A novel approach for multi-label chest X-ray classification of common thorax diseases. IEEE Access **7**, 64279–64288 (2019)
2. Annarumma, M., Withey, S.J., Bakewell, R.J., Pesce, E., Goh, V., Montana, G.: Automated triaging of adult chest radiographs with deep artificial neural networks. Radiology **2018180921** (2019)
3. Baltruschat, I.M., Nickisch, H., Grass, M., et al.: Comparison of deep learning approaches for multi-label chest X-ray classification. Sci. Rep. **9**, 6381 (2019). https://doi.org/10.1038/s41598-019-42294-8
4. Crobu, F., Di Ciaccio, A.: Classify X-ray images using convolutional neural networks. In: Porzio G. C., Greselin F., Balzano S.: CLADAG 2019 Book of Short Papers, pp. 136–139. Centro Editoriale di Ateneo Università di Cassino e del Lazio Meridionale, Cassino (2019)
5. Goodfellow, I., Bengio, Y., Courville, A.: Deep Learning. The MIT Press, Cambridge (2016)
6. Gündel, S., Grbic, S., Georgescu, B., Liu, S., Maier, A., Comaniciu, D.: Learning to recognize abnormalities in chest X-rays with location-aware dense networks. In: Vera-Rodriguez et al. (eds.), Progress in Pattern Recognition, Image Analysis, Computer Vision, and Applications, pp. 757–765. Springer, Cham (2019)
7. He, K., Zhang, X., Ren, S., Sun, J.: Deep residual learning for image recognition. In: 2016 IEEE Conference on Computer Vision and Pattern Recognition (CVPR), pp. 770–778 (2015)
8. Huang, G., Liu, Z., Weinberger, K.Q.: Densely Connected Convolutional Networks. In: 2017 IEEE Conference on Computer Vision and Pattern Recognition (CVPR), pp. 2261-2269 (2016)
9. ImageNet, Large Scale Visual Recognition Challenge (ILSVRC). http://image-net.org/challenges/LSVRC
10. McKinney, S.M., Sieniek, M., Godbole, V., et al.: International evaluation of an AI system for breast cancer screening. Nature **577**, 89–94 (2020). https://doi.org/10.1038/s41586-019-1799-6
11. Wu, N., et al.: Deep neural networks improve radiologists' performance in breast cancer screening. IEEE Trans. Med. Imaging (2019). https://doi.org/10.1109/TMI.2019.2945514
12. Oakden-Rayner, L.: Exploring the ChestXray14 dataset: problems (2017). https://lukeoakdenrayner.wordpress.com/2017/12/18/the-chestxray14-dataset-problems/
13. Rajpurkar, P., et al.: CheXNet: radiologist-level pneumonia detection on chest X-rays with deep learning (2017). arXiv:abs/1711.05225
14. Wang, P., Berzin, T.M., Glissen Brown, J.R., et al.: Real-time automatic detection system increases colonoscopic polyp and adenoma detection rates: a prospective randomised controlled study. Gut **68**, 1813–1819 (2019)
15. Wang, X., Peng, Y., Lu, L., Lu, Z., Bagheri, M., Summers, R.: ChestX-ray8: Hospital-scale chest X-ray database and benchmarks on weakly-supervised classification and localization of common thorax diseases. In: 2017 IEEE Conference on Computer Vision and Pattern Recognition(CVPR), pp. 3462–3471 (2017)
16. Yao, L., Poblenz, E., Dagunts, D., Covington, B., Bernard, D., Lyman, K.: Learning to diagnose from scratch by exploiting dependencies among labels (2017). arXiv:abs/1710.10501

Studying Affiliation Networks Through Cluster CA and Blockmodeling

Daniela D'Ambrosio, Marco Serino, and Giancarlo Ragozini

Abstract In this paper, we propose a new joint approach for analyzing affiliation (two-mode) networks by using factorial methods and blockmodeling. In our recent work, we pursued the joint use of a given factorial method, i.e., MCA/MFA, and a clustering method, namely blockmodeling, but through distinct steps. Here we look for a strategy that permits us to apply the two methods simultaneously. To this aim, we propose a method that allows us to group individuals and variables simultaneously and directly for binary matrices, namely cluster correspondence analysis (cluster CA). This method can be adopted when dealing with affiliation matrices having a binary structure. Hence, we look at the way network positions (clusters) can be incorporated in cluster CA to verify if cluster CA can properly represent specific network structures. We illustrate our proposal through an empirical application on an affiliation network of stage co-productions.

Keywords Affiliation networks · Blockmodeling · Cluster CA · Data classification

1 Introduction

Affiliation networks are a special case of two-mode networks, which consist of two disjoint sets: a set of actors and a set of events in which those actors are involved.

D. D'Ambrosio
Interdepartmental Research Center Urban/Eco, University of Naples Federico II, Naples, Italy
e-mail: daniela.dambrosio2@unina.it

M. Serino (✉) · G. Ragozini
Department of Political Science, University of Naples Federico II, Naples, Italy
e-mail: marco.serino@unina.it

G. Ragozini
e-mail: giragoz@unina.it

S. Balzano et al. (eds.), *Statistical Learning and Modeling in Data Analysis*, Studies in Classification, Data Analysis, and Knowledge Organization,
https://doi.org/10.1007/978-3-030-69944-4_7

One of the main concerns in studying such networks is to establish equivalent classes of actors that are similarly embedded in the whole network, following some criterion of equivalence, such as structural equivalence. Blockmodeling, with its recent extensions [5], permits us to perform a clustering of the affiliation network units.

However, other methods proved equally apt to find relational patterns within affiliation networks. Factorial methods, such as Multiple Correspondence Analysis (MCA) [6], permit us to synthesize, analyze, and graphically represent the relational structure in a metric space. Thanks to the relationships between MCA and blockmodeling, namely the measures that capture structural similarities in a network [3, 4], a joint approach has been proposed to apply a clustering method, i.e., blockmodeling, along with a given factorial method [10, 11].

Nonetheless, in this latter strategy, we can foresee three shortcomings. First, this joint approach is more a tandem analysis than a simultaneous analysis. It performs the different methods through consecutive but distinct steps (sequential approach). Second, clusters do not emerge from the factorial analysis but are added to it only as supplementary variables in order to be projected onto the reduced space. Cluster memberships are thus new categorical variables not involved in the computation of the factorial solution, which only accounts for binary (active) variables representing actors' participation/non-participation in the events. Third, this strategy yields different, albeit compatible, results from the distinct procedures used to analyze the affiliation network structure.

Hence, in this paper, we propose a different method, namely cluster correspondence analysis (cluster CA) [12], which simultaneously groups individuals and variables for binary matrices. It also permits us to evaluate the relations among groups in terms of proper distances in a metric space and with respect to the dimensionality of the factorial solution. By doing that, we attempt at better highlighting and evaluating the relationships among network units both between and within the different clusters (network positions). In line with the main argument of positional network analysis [5], we want to deploy cluster CA to unveil underlying structures (dimensions) in the network data. We present an application of this approach by analyzing the affiliation network of the stage co-productions released in Campania (a region of Italy) during the 2012/2013 season.

2 Factorial Methods and Blockmodeling for Analyzing Affiliation Networks

Recently, a *joint approach* has been proposed that uses MCA and blockmodeling for affiliation networks, relying upon the relationships that exist between factorial methods and blockmodeling. The network positions (i.e., the clusters), as derived from the blockmodeling, are incorporated in the analysis made by MCA as supplementary variables and represented in a metric space [10, 11].

In this approach, clustering and factorial methods, albeit jointly used to analyze the network structure, are kept separated in the analytic process. In this paper, as an advancement of such research line, we propose a factorial method that simultaneously performs a clustering of individuals and variables for binary matrices, the latter being no less than the type of variables concerned with event affiliations (participation or non-participation in a given event).

The method we propose in this work, namely cluster CA, combines cluster analysis and CA and allows us to obtain both a low-dimensional representation of clusters and attributes and a clustering of individuals relying on the profiles related to the categorical variables [12]. Therefore, it permits us to obtain dimension reduction and clustering of categorical data simultaneously [12].

Hence, our main goals are: (i) to look at the way network positions, as they result from blockmodeling, can be incorporated in the cluster CA method and to assess the advantages of this strategy with respect to the one provided by Ragozini et al. [10] (see also [11]); (ii) to analyze specific network structures (e.g., core-periphery and/or segmentation, in line with the theoretical premises discussed in [11]) and to verify if cluster CA can reveal and clearly represent such structures. In fact, we aim to propose this method in that it provides a unique and comprehensive framework for such analytical purposes, in contrast with a sequential approach. The proposed approach will be demonstrated in Sect. 4 by analyzing a real dataset consisting of an affiliation network of stage co-productions.

3 Applying Cluster CA and Blockmodeling to Affiliation Networks

An affiliation network $\mathcal{G}$ can be represented by a triple $\mathcal{G}(V_1, V_2, \mathcal{R})$ composed of two disjoint sets of nodes, V_1 and V_2 of cardinality n and m, and a set of edges or arcs, $\mathcal{R} \subseteq V_1 \times V_2$. By definition $V_1 \cap V_2 = \emptyset$, the two disjoint sets V_1 and V_2 refer to different entities. That is, the set $V_1 = \{a_1, a_2, \ldots, a_n\}$ represents the actor set, whereas the other, $V_2 = \{e_1, e_2, \ldots, e_m\}$, represents the set of m relational events. The edge $r_{ij} = (a_i, e_j)$, $r_{ij} \in \mathcal{R}$, is an ordered couple, and it indicates if an actor a_i attends an event e_j. The set $V_1 \times V_2$ can be fully represented by a binary matrix, the affiliation matrix, $\mathbf{F}(n \times m)$, with element $f_{ij} = 1$ if $(a_i, e_j) \in \mathcal{R}$ and 0 otherwise.

In affiliation networks, the structural equivalence principle states that two actors are equivalent if they participate exactly in the same events [9]. Formally, given two actors a_i and $a_{i'}$, the structural equivalence property $\equiv$ states that $a_i \equiv a_{i'}$ if and only if $r_{ij} = r_{i'j}$ $\forall j$. If two actors a_i and $a_{i'}$ are structurally equivalent, they are indistinguishable, and one equivalent actor can substitute for the other one because the two relational patterns are identical.

To discover the relational structure embedded in $\mathbf{F}$, it is possible to consider it as an usual *case-by-variable* matrix and then apply a factorial method like MCA. In applying the latter, the indicator matrix $\mathbf{Z}$ is derived from the matrix $\mathbf{F}$ through full disjunctive coding. Given that each relational event e_j is a dichotomous variable, the indicator matrix $\mathbf{Z}$ contains two columns for each e_j, namely e_j^+ and e_j^-, where e_j^+ is the value of a dummy variable coding the participation in the event, and e_j^- is the value of a dummy variable coding the non-participation. As all the variables in $\mathbf{F}$ are dichotomous, the corresponding indicator matrix $\mathbf{Z}$ turns out to be a *doubled matrix*.

Given our affiliation matrix **F** and the (doubled) indicator matrix **Z** derived from the former, and following the approach proposed by van de Velden et al. [12], we aim to find $\mathbf{Z}_K$. It is the indicator matrix of dimensionality $n \times K$, which includes the cluster membership considered as a categorical variable such that $\mathbf{F}^c = \mathbf{Z}'_K\mathbf{Z}$ is the table of cross-tabulations that includes all the associations between the cluster memberships and the binary variables coding the participation (and non-participation) in events.

Following the iterative procedure described by van de Velden et al. [12], skipping its technical details, we propose to apply the algorithm for cluster CA as follows:

1. generate an initial cluster allocation $\mathbf{Z}_K$;
2. find category quantifications by using the usual CA algorithm;
3. construct an initial configuration of the relational patterns for the actors **Y** (as defined by van de Velden et al. [12]);
4. update the membership matrix $\mathbf{Z}_K$ by applying a clustering method to **Y**, and
5. repeat the procedure (i.e., go back to step 2) until convergence.

In the original paper [12], the first solution that was proposed was the random assignment, while the clustering algorithm was the k-means. In this paper, we compare the performance of such methods with the use of blockmodeling to provide both the initial cluster allocation $\mathbf{Z}_K$ and its updating. In this way, the network positions should be optimally separated with respect to the distributions over the events and, simultaneously, events with different participation patterns should be optimally separated [12].

The proposed approach will be demonstrated in the next section by analyzing a real dataset consisting of an affiliation network of stage co-productions.

4 A Case Study of Stage Co-productions

The affiliation network that we analyze in this paper stems from a wider dataset of theater companies and co-productions which we dealt with in our previous work [11]. The original dataset comprised the stage co-productions that 20 theater companies located in the Campania region of Italy released with 80 other companies over four theater seasons (from 2011 to 2015). In this work we focus on the 43 co-productions (i.e., the events in the affiliation network) released and performed in that region during the 2012/2013 season, which involved 40 companies (i.e., the actors). We collected data on the co-productions and the companies involved in them by means of web-based questionnaires completed by companies' staff, along with those companies' websites to complement the data (for more details see [11]). In this data structure, where the rows represent the companies and the columns represent the stage co-productions, we expect to find groups of theater companies that share similar participation patterns and that are involved in co-productions with similar characteristics (i.e., belonging to the same artistic genres). At the same time, we attempt to evaluate the structural similarities between the groups of companies on the basis of their projections in a metric space.

Table 1 Global results of the Cluster CA solutions from different ways to start initial cluster allocation: randomly and by blockmodeling positions

	Random start	Blockmodeling positions
Number of clusters of the best solution	5	6
Number of dimensions of the best solution	4	5
Average silhouette width	0.57	0.12
Objective criterion	20.63	21.46
Between-SS/Total-SS	0.94	0.91

Participation in this kind of collaboration can be done for specific reasons. Indeed, stage co-productions are intended to share costs and gain mutual advantages thanks to the optimization of financial, material, and human resources and to pursue a joint artistic project by sharing aesthetic and socio-cultural views.

As shown in our previous research [11], we know that the relational patterns of co-productions can be more or less segmented. The motives for co-producing plays (e.g., personal relations or preferred artistic genres) may induce producers to enter more or less exclusive alliances with several partners. Furthermore, companies with interdependent economic or symbolic resources, or with similar artistic motivation, will participate in co-productions in relative isolation from other partners.

As noted above, we analyze the co-production network structure by means of cluster CA, comparing its results with those obtained by the blockmodeling procedure on the same network and using the network positions (clusters) that derive from the blockmodeling as a custom starting partition for cluster CA.

4.1 Comparison of Different Ways to Start Initial Cluster Allocation: Randomly and by Blockmodeling Positions

We run cluster CA using the ‘clustrd’ R package [7, 8], and we analyze the co-production network in two ways: using the random procedure and the blockmodeling positions (according to our previous work [11]) to start clustering. The results are illustrated in Tables 1, 2, and 3, where the performances of the two different procedures of the method are compared.

In both ways, the first step is to identify the proper number of clusters and dimensions to be extracted. We make this choice through a procedure implemented in the ‘tuneclus()’ function of the ‘clustrd’ R package [7, 8]. It facilitates the selection of the appropriate number of clusters and dimensions for the joint dimension reduction. It also helps with selecting the clustering methods by assessing the cluster’s quality for a certain range of clusters and dimensions. We select the solutions with an optimal number of dimensions and clusters based on the average silhouette width (ASW) index, which ranges from –1 to 1. The ASW index reflects the compactness of the

Table 2 Results (by cluster) of the cluster CA solutions from different ways to start initial cluster allocation: randomly and by blockmodeling positions

	Size	% Tot.	ASW	Within cluster sum of squares by cluster
CA clusters and members	Cluster CA with random start			
Cluster 1 - Other	32	80%	0.60	2.15
Cluster 2 - CampaniadeiFestival/Stabile Napoli	2	5%	0.30	1.34
Cluster 3 - Bracco/Totò	2	5%	0.18	0.04
Cluster 4 - TeatrAzione/Magazzini	2	5%	0.85	0.63
Cluster 5 - Progetto Museo/Nuvole	2	5%	0.41	1.76
CA clusters and members	Cluster CA with blockmodeling positions			
Cluster 1 - Ex-Asilo and other	18	45%	–0.09	2.85
Cluster 2 - Other	13	32.5%	0.30	0.67
Cluster 3 - TeatrAzione/Magazzini/Carrozza	3	7.5%	0.03	3.63
Cluster 4 - Progetto Museo/Nuvole	2	5%	0.13	2.19
Cluster 5 - Bracco/Totò	2	5%	0.85	0.07
Cluster 6 - CampaniadeiFestival/Stabile Napoli	2	5%	0.23	1.82

clusters, and it indicates if the cluster structure is well separated (values near 1) or not (values near –1) [8]. As shown in Tables 1 and 2, the random start procedure achieves a (slightly) higher performance than the one based on the blockmodeling positions, where the optimal ASW value is 0.57 and is obtained for a solution with 5 clusters and 4 dimensions. Furthermore, as shown in Table 3 and on the factorial plane in Fig. 1, the clustering results of the two procedures of cluster CA are similar to each other and also to those that derive from the blockmodeling classification. In both cases, the best solution reproduces the classification we found for this network, with 6 or 5 clusters (starting from the blockmodeling or randomly, respectively). In particular, the solution obtained through random start turns out to parallel the one obtained by MCA (with the advantage of performing simultaneous classification and dimension reduction). The differences between the results of the two clustering methods (i.e., blockmodeling and cluster CA) shown in Table 3 seem to reflect the ability of the cluster CA method to identify only non-random patterns of structures.

These results allow us to identify three levels of meaning from three perspectives. Notably, the location of the points on the factorial plane and their distances from the axes origin can be read in terms of the following:

1. Network topology from the SNA and graph theory perspectives: points that lie close to the origin indicate structures tending to random graphs; conversely, points that are located far from the origin indicate the presence of some structure other than a random graph.

Table 3 Comparison of clustering results: blockmodeling and the two cluster CA procedures

		Blockmodeling positions and members					
		C1-PN	C2-BT	C3-CS	C4-TMC	C5-Ex-A/O	C6-O
Clustering of the cluster CA	By random start						
	Cluster 1-Other	0	0	0	1	18	13
	Cluster 2-CS	0	0	2	0	0	0
	Cluster 3-BT	0	2	0	0	0	0
	Cluster 4-TM	0	0	0	2	0	0
	Cluster 5-PN	2	0	0	0	0	0
	By block-modeling positions						
	Cluster 1-Ex-A/Other	0	0	0	0	18	0
	Cluster 2 - Other	0	0	0	0	0	13
	Cluster 3 - TMC	0	0	0	3	0	0
	Cluster 4 - PN	2	0	0	0	0	0
	Cluster 5 - BT	0	2	0	0	0	0
	Cluster 6 - CS	0	0	2	0	0	0

Note: columns show blockmodeling positions (clusters); rows indicate clusters of the cluster CA solution (random above and by blockmodeling positions below)

2. Distance from the independence hypothesis in the statistical perspective: points that lie close to the origin show structures that approximate the independence condition, while points that are located far from the origin show structures and patterns that are far from the independence condition.
3. Substantive meaning in a theoretical perspective: points that lie close to the origin indicate unstable and unstructured relational patterns, namely occasional collaborations, whereas points that are located far from the origin suggest stable and structured relational patterns, namely habitual and strong collaborations.

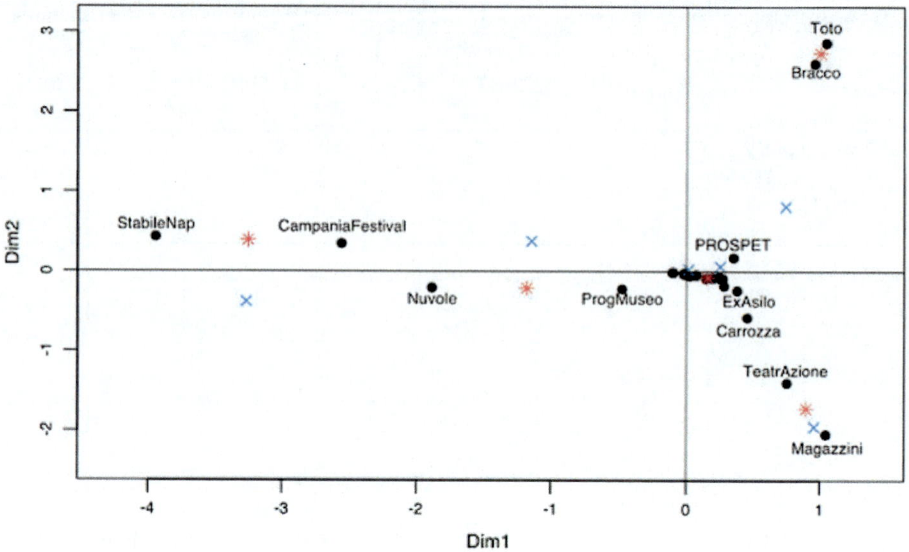

Fig. 1 Representation of organizations and Random Start (red stars) and Blockmodeling Positions Start (blue crosses) centroids of the clusters in the space of co-productions. Labels of points close to the center are omitted

4.2 *Cluster CA for Affiliation Networks: Results From Random Start Clustering*

As noted, in this case, the results of the two procedures are very similar. However, we choose to show only those related to the random start version, as they turned out to be slightly better in discriminating between the most stable and significant relational patterns and those without a clear structure (casual and unstable collaborations). In fact, cluster CA with a random start procedure seems able to identify non-random association patterns in a clearer way.

Hence, focusing on the random start solution (Fig. 2), we note that companies belonging to the clusters that are located far from the origin—namely Clusters 2, 3, 4, and 5—are characterized by specific patterns of participation in co-productions, those patterns being very distant from the probability of independence. Instead, Cluster 1 lies close to the origin of the axes in that the related companies have a common participation profile. That is to say, the profile is close to the null hypothesis (no association between the participation profiles of those companies). For those companies, there is no clearly identifiable pattern (they participate "by chance" in co-productions). Figure 3 shows the co-productions for each cluster that deviate most from the independence condition. This allows us to provide the reader with further comments regarding the composition of the clusters and the co-productions characterizing them.

On the basis of our prior knowledge of the field under investigation [10, 11], we then attempt to interpret the joint representation shown in Fig. 2 by examining

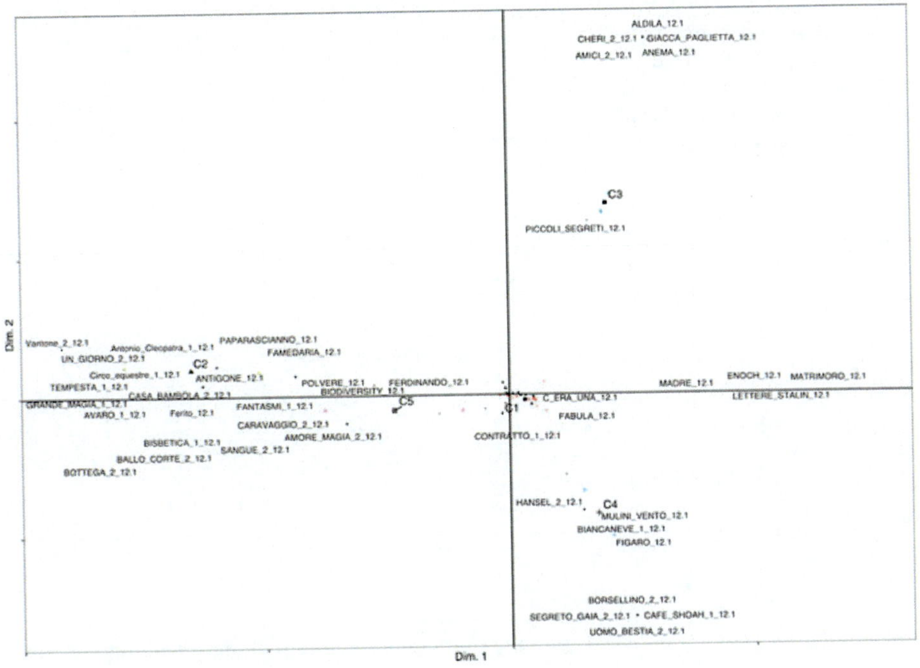

Fig. 2 Joint representation of organizations, centroids of the clusters, and co-productions

the positions of the cluster centroids and the co-productions on the factorial planes. The first dimension discriminates between clusters and co-productions that denote higher amounts of cultural (symbolic) capital [1, 2]. On the left-hand side of the map, Clusters 2 and 5 are made of distinguished organizations that have a clear orientation toward the dissemination of "high culture". On the right-hand side are the clusters and co-productions with less prestige and/or cultural vocation and that are more bent on entertainment, namely Cluster 3 and 4. More specifically, organizations belonging to the clusters located on the left-hand side benefit from formal acknowledgment by the state or local authorities as key institutions in the cultural field [1], as is the case with publicly funded theaters like Stabile Napoli (Cluster 2) and Nuvole (Cluster 5). This is also the case for Campania dei Festival (Cluster 2), a foundation that, thanks to public subsidies, manages a renowned theater festival. In fact, this festival guarantees accrued symbolic capital to the theater companies and performers that appear in its program [11].

Another organization, namely Progetto Museo (Cluster 5), is an association of art historians devoted to promoting educational activities related to the cultural heritage, and to museums in particular, in the Campania region. Liaisons - i.e., strong relationships based on recurrent co-productions that reinforce each other's cultural capital - occur between the companies that belong to each of these two clusters. The same culturally oriented purposes hold true for the co-productions located on the same side of the map. They comprise well-known classic and contemporary

Fig. 3 Co-productions that deviate the most from the independence condition (C1 has been removed because it is not informative)

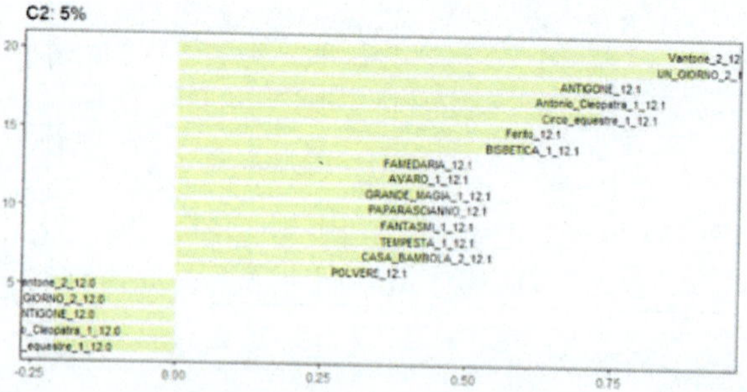

(a) Cluster 2: Campania dei Festival - Stabile Napoli

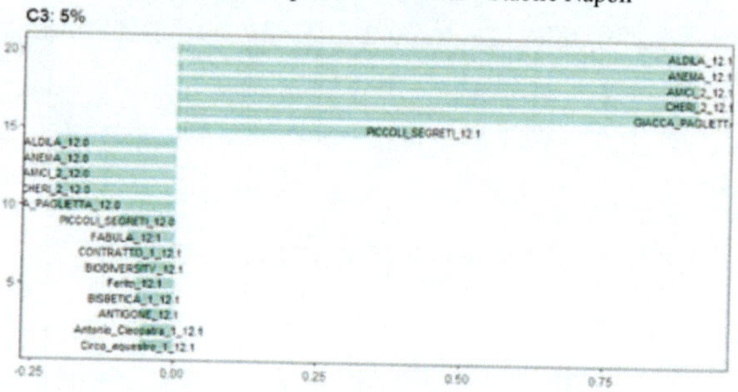

(b) Cluster 3: Bracco-Totò

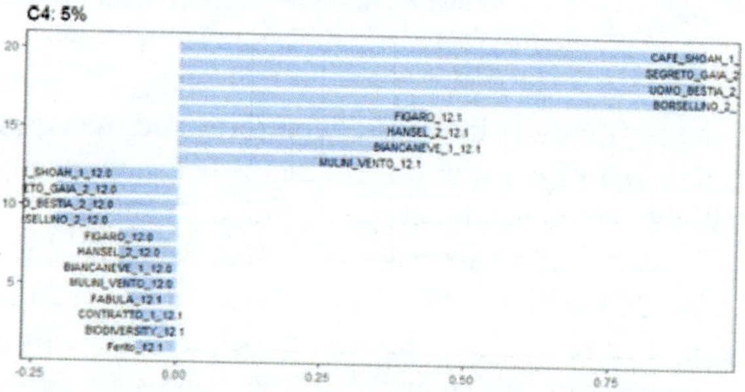

(c) Cluster 4: TeatrAzione - Magazzini

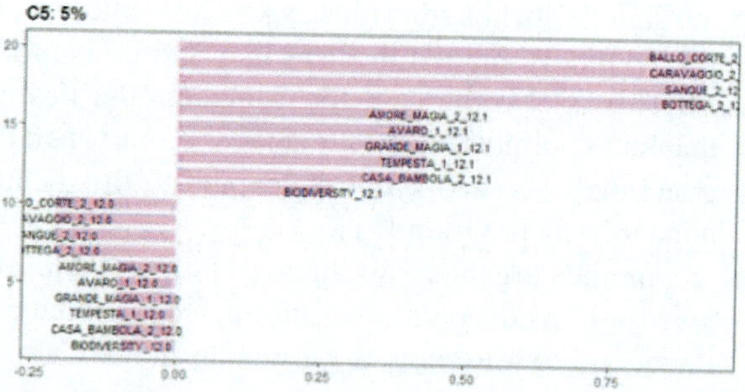

(d) Cluster 5: Progetto Museo - Nuvole

plays often presented with revised play scripts, but also youth theater or new drama projects, with a key propensity for more serious themes and elaborate styles and languages. These traits are clearly present in co-productions that most deviate from the independence condition in Cluster 2, in Fig. 3a. They are Plauto's *Vantone*, the innovative drama project *Un giorno tutto questo sarà tuo* (denoted by the label *UN GIORNO*), a revisited version of *Antigone*, and Shakespeare's *Antonio e Cleopatra*, which can also be seen on the left-hand side of the factorial plane (Fig. 2). As for Cluster 5(d), the labels *BALLO CORTE*, *CARAVAGGIO*, *SANGUE* and *BOTTEGA* denote co-productions aimed at joining theater, history, art, and education. They confirm what was said above by their marked deviation from the independence condition (see also the left-hand side of the map in Fig. 2).

The right-hand side of the map is instead characterized by the presence of the companies Bracco and Totò (Cluster 3), which devote themselves to the traditional and markedly popular segment of Neapolitan dialect productions made by local playwrights and directors. Those productions are, nonetheless, very different from the plays by Eduardo De Filippo, a renowned and "cultured" figure of the Neapolitan theater, these plays being often presented by the companies of the Clusters 2 and 5. In fact, the five co-productions that are more distant from the independence condition in Cluster 3, as shown in Fig. 3b, are all Neapolitan comedies and musicals of the kind mentioned above. However, on the bottom-right of the map in Fig. 2, Magazzini and TeatrAzione (Cluster 4) are, instead, more steadily devoted to youth theater, but they lack the amount of symbolic recognition that characterizes the companies of Clusters 2 and 5. In fact, differently from the latter, they are not acknowledged by the state as key theater institutions. Indeed, the companies Magazzini and TeatrAzione put great effort into producing plays that convey ideas about socially relevant issues which suit the educational needs of the youth and of school pupils. This is the case with the four co-productions (*CAFE SHOAH*, *SEGRETO GAIA*, *UOMO BESTIA*, *BORSELLINO*) more distant from the probability of independence in Cluster 4 (see Fig. 3c and Fig. 2, bottom-right). However, they do not receive the same recognition for their works as that granted to the productions in Clusters 2 and 5.

The second dimension best highlights the opposition between educational and entertainment purposes. It clearly distinguishes the companies belonging to Cluster 4—but also to those in Cluster 5 (in that companies like Nuvole and Magazzini share a preference for educationally driven pieces)—from the mere entertainment that characterizes the companies of Cluster 3. These latter are more clearly market-driven and prone to respond to the audience's needs for amusement and "popular" theater pieces (e.g., comedy, cabaret, and musicals).

5 Concluding Remarks

By focusing at the same time on classification and dimension reduction, cluster CA helps us understand how and why we observe certain relational (structural) patterns in an affiliation network. Given the results we have obtained, this method may be a good candidate to accompany blockmodeling in the analysis of affiliation networks.

In sum, we noted that cluster CA allowed us to both synthesize the relevant network structure and obtain a classification of actors starting from their event affiliations. Moreover, it provided a joint representation, in a metric space, of both actors/events and the cluster they belong to, allowing us to quantify their distances.

In conclusion, we believe that this method can be further developed as follows. First, we need to test the cluster CA method with more complex network structures. Second, it can be useful to perform a simulation study to evaluate the performances of the cluster CA method with network structures that have different characteristics (density, etc.). Finally, in an SNA perspective, it may be interesting to find a way to modify the criterion function of the cluster CA so that the performed procedure leads to indirect blockmodeling.

References

1. Bourdieu, P.: The field of cultural production, or: the economic world reversed. Poetics **12**(4), 311–356 (1983)
2. Bourdieu, P.: The forms of capital. In: Richardson, J.G. (ed.) Handbook of Theory and Research for the Sociology of Education, pp. 241–258. Greenwood Press, New York (1986)
3. D'Esposito, M.R., De Stefano, D., Ragozini, G.: A Comparison of χ^2 Metrics for the Assessment of Relational Similarities in Affiliation Networks. In: Vicari, D., Okada, A., Ragozini, G., Weihs, C. (eds.) Analysis and Modeling of Complex Data in Behavioral and Social Sciences, pp. 113–122. Springer, Cham, Switzerland (2014)
4. D'Esposito, M.R., De Stefano, D., Ragozini, G.: On the use of multiple correspondence analysis to visually explore affiliation networks. Soc. Netw. **38**, 28–40 (2014)
5. Doreian, P., Batagelj, V., Ferligoj, A.: Generalized Blockmodeling. Cambridge University Press, Cambridge (2005)
6. Greenacre, M., Blasius, J.: Multiple Correspondence Analysis and Related Methods. CRC Press, Boca Raton (2006)
7. Markos, A., Iodice D'Enza, A., van de Velden, M.: Clustrd: Methods for Joint Dimension Reduction and Clustering. R package version 1.3.6-2 (2019). https://CRAN.R-project.org/package=clustrd
8. Markos, A., Iodice D'Enza, A., van de Velden, M.: Beyond tandem analysis: joint dimension reduction and clustering in R. J. Stat. Softw. **91**(10), 1–24 (2019). https://doi.org/10.18637/jss.v091.i10
9. Pizarro, N.: Structural identity and equivalence of individuals in social networks: beyond duality. Int. Sociol. **22**(6), 767–792 (2007)
10. Ragozini, G., Serino, M., D'Ambrosio, D.: On the analysis of time-varying affiliation networks: the case of stage coproductions. In: Perna, C., Pratesi, M., Ruiz-Gazen, A. (eds.) Studies in Theoretical and Applied Statistics. SIS 2016, Salerno, Italy, June 8–10, pp. 119–129. Springer, Cham, Switzerland (2018)
11. Serino, M., D'Ambrosio, D., Ragozini, G.: Bridging social network analysis and field theory through multidimensional data analysis: the case of the theatrical field. Poetics **62**, 66–80 (2017)
12. van de Velden, M., Iodice D'Enza, A., Palumbo, F.: Cluster correspondence analysis. Psychometrika **82**(1), 158–185 (2017)

Sectioning Procedure on Geostatistical Indices Series of Pavement Road Profiles

Mauro D'Apuzzo, Rose-Line Spacagna, Azzurra Evangelisti, Daniela Santilli, and Vittorio Nicolosi

Abstract Road sectioning plays a crucial role in Road Asset Management Systems and nowadays high-speed laser-based devices are able to quickly collect a huge amount of data on pavement surface characteristics. However, collected data cannot be directly employed in road maintenance planning but synthetic values have to be derived and this implies a high computational effort in identifying effective synthetic indices and road homogeneous sections. To this purpose, the Geostatistical tools, in terms of Variogram scheme have been applied for characterizing road surface. "Range" and "Sill" values, deriving from the Variogram application, have been proposed as pavement surface characteristics synthetic indices (namely the macrotexture) to identify different road surfaces. Once that Variogram scheme has been applied, a dynamic sectioning procedure can be employed to detect homogeneous road pavement sections and compared with more traditional descriptors. Preliminary results obtained by an experimental smart road, seem to highlight that the Variogram variables can be promising in both road texture characterization and homogeneous section identification.

M. D'Apuzzo · R.-L. Spacagna · A. Evangelisti (✉) · D. Santilli
Department of Civil and Mechanical Engineering (DICeM), University of Cassino, Cassino, Italy
e-mail: aevangelisti.ing@gmail.com

M. D'Apuzzo
e-mail: dapuzzo@unicas.it

R.-L. Spacagna
e-mail: rlspacagna@unicas.it

D. Santilli
e-mail: daniela.santilli@unicas.it

V. Nicolosi
Department of Enterprise Engineering "Mario Lucertini", University of Rome "Tor Vergata", Rome, Italy
e-mail: nicolosi@uniroma2.it

S. Balzano et al. (eds.), *Statistical Learning and Modeling in Data Analysis*, Studies in Classification, Data Analysis, and Knowledge Organization,
https://doi.org/10.1007/978-3-030-69944-4_8

Keywords Pavement management · Road surface macrotexture · Geostatistics variogram scheme · Spatial data analysis · Dynamic sectioning

1 Introduction

A Pavement Management System (PMS) is a decision support system that provides effective help to road managers for planning maintenance interventions on a road pavement network in order to reach predefined performance goals that are consistent with budget constraints within a short-, medium-, and long-term scenarios. Nowadays, a PMS acts as a "sub-module" of a more general Transportation Infrastructure Asset Management System comprising all the different facilities (such as safety barriers, lighting or hydraulic systems, geotechnical structures, and so on). However, its basic principles rely on the possibility to describe pavement condition by means of several parameters that can be measured and collected along the road on a routine basis. Among the different parameters collected according to the existing Road Standards and Guidelines [2, 3], pavement surface characteristics [19] are the most significant as they affect several functional properties of road pavements such as the vehicle riding comfort and the tire-road friction, noise, and rolling resistance. Pavement surface characteristics are mainly measured by spectrally decomposing the acquired longitudinal road profiles in order to evaluate the different texture scales. The macrotexture scale, which is associated with the road profile wavelengths lying between the 0.5 and 50 mm range [7, 8, 11, 19], appears to be one of the most critical as it affects skid resistance, splash and spray phenomenon, hydroplaning, tire-pavement noise, and rolling resistance. As a matter of fact, different macrotexture descriptive indices can be derived: one of the most known and used is an indirect measure called Mean Profile Depth (MPD) evaluated according to [2], although more reliable macrotexture synthetic indices have been recently proposed [7] together with new methods for the texture prediction [8]. On the other hand, in the past decades, measuring methods and techniques for pavement condition evaluation made great strides and several High-Speed Laser-based (HSL) measuring devices have been developed and employed. Due to technologies, operating conditions, and intrinsic heterogeneous nature of the road pavement surface, the one-dimensional sampled profile can be affected by noise and invalid readings. For these reasons the HSL data usually undergo a pre-processing (filtering) procedure, according to several approaches [7, 16]. However, the huge amount of data collected cannot be directly used in the databases, but must be previously analyzed, in order to identify the statistically significant values to be associated with each homogeneous road section (pavement condition parameters as almost constant). Several sectioning methods are available in the literature; however, different aspects have to be still investigated in order to identify better synthetic indices for the texture characterization. In this paper, an innovative approach is proposed to describe the macrotexture of road surface employing geostatistical tools for characterizing one-dimensional road profiles. Then the sectioning process has been applied on both traditional synthetic index

(MPD) and spatial index obtained by the geostatistical approach showing a better performance of the latter proposed approach.

2 Geostatistical Tools for Road Pavement Characterization

Geostatistics is a field of the Statistics focused on the study of spatial or regionalized phenomena, which are characterized by a spatial correlation [5]. Thanks to this peculiarity, several applications within environmental aspects have been performed [21, 22] and encouraging results have been achieved from preliminary attempts for the road profiles analysis [10]. In this case, the spatial structure of the pavement texture has been studied using the geostatistical tools, highlighting a correlation between the pavement characteristics (grain size and binder) and the spatial properties of the Variogram. This spatial tool describes the relation between two measured point at "h" distance. The experimental variogram $\gamma(h)$ is estimated following Eq. 1:

$$\gamma(h) = \frac{1}{2N(h)} \sum_{i=1}^{N(h)} (Z(x_i + h) - Z(x_i))^2 \tag{1}$$

where $Z(x_i)$ is the measured variable at location x_i, $Z(x_i+h)$ is the measured variable at location x_i+h, and N(h) is the number of couple of points at h distance. The spatial structure is obtained by modeling the experimental variogram according to the specific functions proposed in literature [5, 18]. The most common simple models are reported in Table 1.

Those variogram models are increasing function with distance, characterized by two properties: the range "a" and the sill "C". The range represents the distance beyond which the data exhibit a spatial correlation, and the sill is the value of the variogram reached the range. Several methodologies are available to automatically fit the chosen model with the experimental variogram. The optimal definition of the

Table 1 The most common simple variogram models

Model	Equation	
Spherical	$\gamma(h) = \begin{cases} C\left[\left(\frac{3h}{2a}\right) - \left(\frac{h^3}{2a^3}\right)\right], & for \quad 0 \le h < a \\ C, & for \quad h \ge a \end{cases}$	(2)
Exponential	$\gamma(h) = C\left[1 - e^{\left(-\frac{3h}{a}\right)}\right]$	(3)
Gaussian	$\gamma(h) = C\left[1 - e^{\left(-3\frac{h^2}{a^2}\right)}\right]$	(4)
Cubic	$\gamma(h) = \begin{cases} C\left[7\left(\frac{h}{a}\right)^2 - \frac{35}{4}\left(\frac{h}{a}\right)^3 + \frac{7}{2}\left(\frac{h}{a}\right)^5 - \frac{3}{4}\left(\frac{h}{a}\right)^7\right], & for \quad 0 \le h < a \\ C, \quad for \quad h \ge a \end{cases}$	(5)

sill and the range can be obtained by using standard minimization procedures [18]. In this paper, the calculations have been performed with R software [20] and RGeostats plugin [17]. The automatic variogram modeling is based on an iterative least square algorithm, called foxleg [9]. Both MPD and the variogram model properties, are considered as indices of the pavement texture and used to identify the homogeneous pavement sections.

3 Brief Overview on Road Sectioning Methods

Pavement condition data usually vary along the road alignment and since data sampling has always been pursued by means of a discrete approach sectioning method were borrowed by typical industrial process control techniques. The key aspect is based on the identification of the transition point between two adjacent homogeneous sections, namely the "break points". Break points can be detected by means of graphical or statistical approach. The Cumulative Difference Method (CDA) proposed by the American Association of State Highway Transportation Officials (AASHTO) [1] and the method of CUMulative SUMs (CUMSUM) [4] use a graphical approach and became very popular because of the ease of implementation and use in PMS, however, this approach gives rise to some problems related to objective identification of transition points. Statistical approaches offer more sound and automated methods to identify the position of breakpoints since their basic principle relies on the fact that pavement data collected can be described as time series characterized by structural changes. These latter methods can be further distinguished in the following:

- linear, if the algorithm to detect and statistically verify the break points is sequentially applied along the road chainage, thus by an analysis approach based on a moving window,
- non-linear, if the method is applied to the entire dataset thus providing the optimal partition that meets predefined requirements and statistical criteria.

In the former group, the most significant are the Dichotomic method developed by Laboratoire Central du Ponts et Chaussées (LCPC) [15] and the Pruned Exact Linear Time with the Empirical Distribution of the cost function (ED-PELT) [12], whereas at the latter one belong: the Bayesian Methods [23], minimum Root Mean Square (RMS) based methods (MINRMS), or the Linear Model with Multiple Structural Changes (LMSC) Method as the method introduced by James and Matteson [13] and the method developed by Killick, Fearnhead, and Eckley [14]. A benchmarking among these different methods is reported in [6].

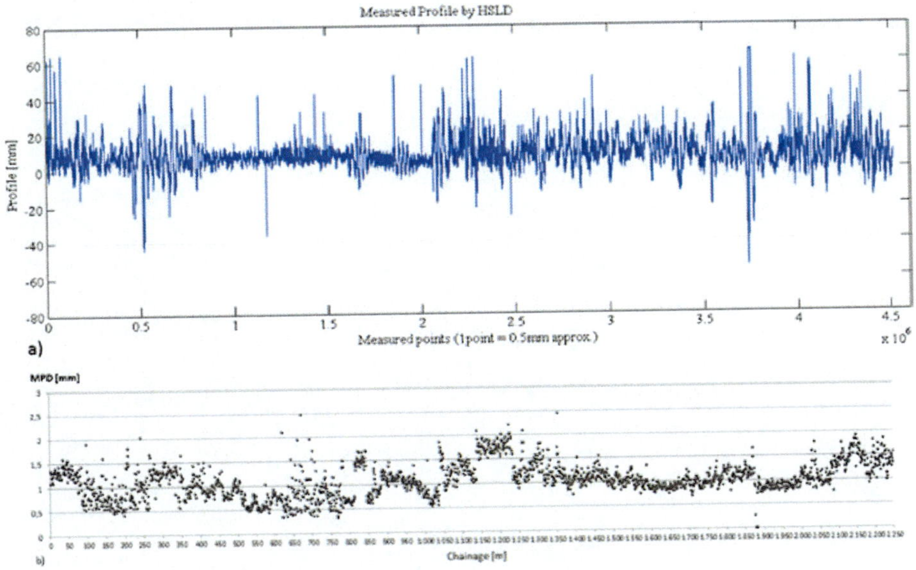

Fig. 1 **a** Measured profile by the HSL and **b** corresponding MPD values along the road profile

4 Data Collection and Analysis

The road pavement characterization is mainly based on the analysis of the road surface, which can be sampled on a one-dimensional or two-dimensional basis. In this paper, a single one-dimensional road profile has been collected with sample spacing of about 0.5 mm (see Fig. 1a) by means of an HSL device, with a laser spot of 0.2 mm and a sampling frequency of 64 kHz. Pavement profile measurement has been performed at the Virginia Smart Road, which is a full-scale, closed test-bed research facility managed by the Virginia Tech Transportation Institute (VTTI), where 24 different road pavement typologies have been laid along an overall length of about 2300 m. According to the [2], the sampled profile underwent a cleaning process, enabling to remove spikes, drop-outs, and data trend. On the filtered profile in parallel the MPD has been evaluated (see Fig. 1b) and the Geostatistical tools have been applied. Due to the one-directional sampling of the collected road profiles, the isotropy and anisotropy analysis has been inevitably reconducted to the evaluation of one-directional experimental semivariograms.

The one-directional experimental semivariograms of the profile have been calculated with a lag distance of 0.5 mm and the number of lag of 30 (15 mm) on a moving window of 1 m. The four models of Table 1 have been automatically fitted (Fig. 2).

In order to identify the appropriate model, statistical tests in term of the Pearson correlation coefficient (ρ), the Kendall rank correlation coefficient (τ), the Spearman's rank correlation coefficient (r_s), and the R^2 (also in term of angular coefficient and intercept), have been calculated and summarized in Table 2. As can be seen, the

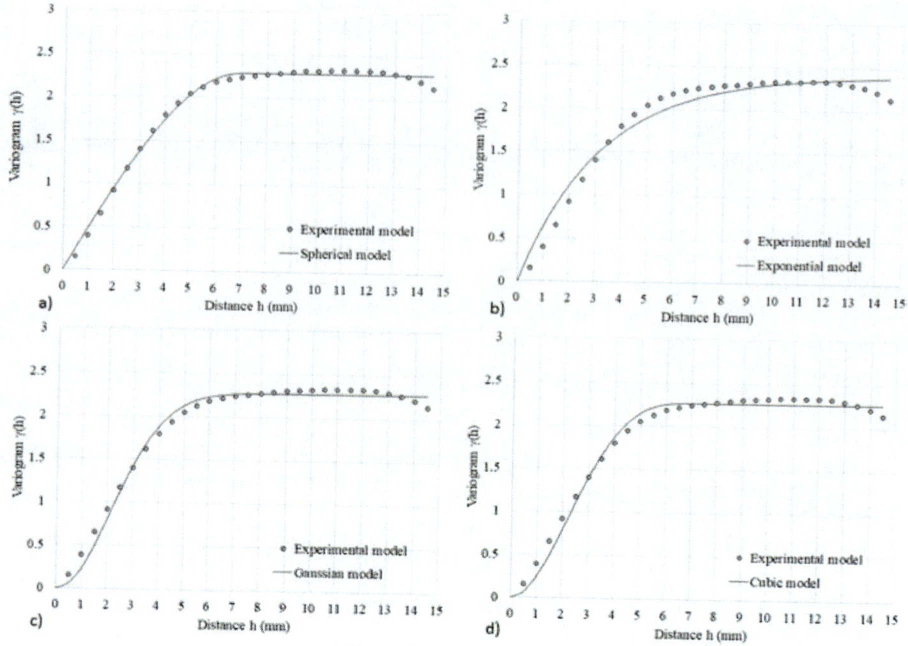

Fig. 2 Experimental semivariogram of a meter of road pavement and **a** Spherical, **b** Exponential, **c** Gaussian and **d** Cubic variogram Models

Table 2 Average values of the statistical descriptors for comparison between the different variogram models

Model	ρ	τ	r_s	R^2	Ang. Coeff.	Intercept
Spherical-Fig. 2a	**0.9880**	**0.7869**	**0.8660**	**0.9763**	**0.9756**	**0.0004**
Exponential-Fig. 2b	0.9877	0.7041	0.8038	0.9757	1.0342	−0.0410
Gaussian-Fig. 2c	0.9787	0.7071	0.8015	0.9582	0.9176	0.0491
Cubic-Fig. 2d	0.9780	0.7737	0.8566	0.9568	0.9166	0.0496

Spherical Model is more appropriate to describe the experimental semivariogram evaluated on the road profiles.

The Range and Sill evaluated by the Spherical model have been represented in Fig. 3. As it is possible to see, the Range and Sill (R&S) series describe two different features of the same measured profile thus providing additional information on structural changes that can be used by sectioning methods.

In order to perform the sectioning process, among the aforementioned statistical methods, the LCPC Method, the ED-PELT, and the methods proposed in [13, 14], on both MPD and the R&S series, have been employed and compared. In this case, the comparison in terms of the number of identified real break points has been summarized in Table 3.

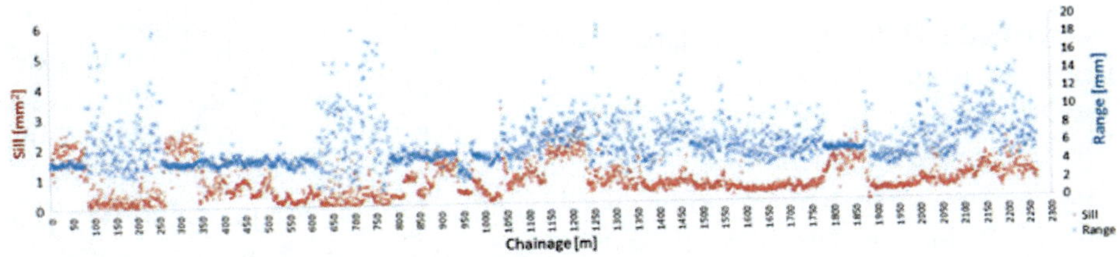

Fig. 3 Sill and range representation along the road profile

Table 3 Identification of the homogeneous segments with different sectioning methods on both MPD and R&S series

Sectioning methods	Series	
	R&S	MPD
ENVCPT package—mean	17/24	15/24
ENVCPT package—AR1	17/24	8/24
ENVCPT package—AR2	16/24	7/24
CHANGEPOIT.NP package (mean)	17/24	13/24
ECP package	11/24	15/24
LCPC	**20/24**	**15/24**

The LCPC method, with a significance level (α) = 5% and sample size of 25, appears to be the more efficient in the identification of the homogeneous pavement road sections, moreover, the R&S indices provide more satisfactory results than the MPD for 5 sectioning methods on 6 tested.

5 Conclusion

In Pavement Management Systems, pavement condition data are nowadays collected by means of high-performance measuring devices. However, the acquired huge amount of data requires a sectioning analysis in order to obtain synthetic descriptors to be used in the planning of maintenance interventions for specific road sections. The basic idea is to apply a Variogram scheme, derived from the Geostatistics field, to the filtered road profile in order to obtain a transformed dataset that is characterized by two statistical descriptors (Range and Sill—R&S). It is believed that this dual representation can better highlight the structural changes in the dataset in order to improve the effectiveness of road sectioning procedures, with respect to more traditional descriptors such as the MPD. A preliminary experimental validation of this procedure has been carried out on real pavement data collected by means of the HSL Device on the Virginia Smart Road (an experimental track that is composed of several pavement sections). Different variogram models have been applied and compared and the Spherical one was more appropriate to describe the road texture; its relative R&S indices have been selected, together with the MPD, for the following sectioning process. In order to detect homogeneous road sections, different methods have been used and compared. The obtained results showed that the R&S indices produce more satisfactory results than MPD (for 5 sectioning procedures on 6) and the LCPC method detects the highest number of real breaks. However, further investigations have to be carried out in order to improve the proposed procedure and to extend the validation to a wider dataset.

Acknowledgements The authors would like to thank Prof. Gerardo Flintsch and the staff of the Virginia Tech Transportation Institute for providing pavement profile data.

References

1. AASHTO: Guide for design of pavement structures, American Association of State Highway and Transportation Officials, Washington D.C. (1986)
2. ASTM E1845-15: Standard Practice for Calculating Pavement Macrotexture Mean Profile Depth, American Society for Testing and Materials (ASTM) International, West Conshohocken, PA (2015). www.astm.org. https://doi.org/10.1520/E1845-15
3. ASTM E965: Standard Test Method for Measuring Pavement Macrotexture Depth Using a Volumetric Technique. American Society for Testing and Materials (2006). https://doi.org/10.1520/E0965-96R06
4. Austroad: Consistency in approaches to road network segmentation and data aggregation – review of current practice. Austroad pubblication No. AP-R276/05, Austroad Inc., Sydney (2005). ISBN 1921139072
5. Chilès, J.P., Delfiner, P.: Geostatistics: Modeling Spatial Uncertainty. Wiley, New York (1999)
6. D'Apuzzo, M., Nicolosi, V.: Detecting Homogeneous Pavement Section Using Econometric Test for Structural Changes in Linear Model. Transportation Research Board 91st Annual Meeting Paper no 12-2125, 0–18, Transportation Research Board, Washington DC, United States (2012)
7. D'Apuzzo, M., Evangelisti, A., Flintsch, G.W., de Leon Izeppi, E., Mogrovejo, D.E., Nicolosi, V.: Evaluation of Variability of Macrotexture Measurement with Different Laser-Based Devices. Airfield and Highway Pavements: Innovative and Cost-Effective Pavements for a Sustainable Future. 294-305. TRIS, ASCE (2015). https://doi.org/10.1061/9780784479216.027
8. D'Apuzzo, M., Evangelisti, A., Nicolosi, V.: Preliminary investigation on a numerical approach for the evaluation of road macrotexture, vol. 10405, pp. 157–172 (2017). https://doi.org/10.1007/978-3-319-62395-8. In Lecture Notes in Computer Science (Including Lecture Notes in Artificial Intelligence and in Bioinformatics)-ISBN:978-3-319-62394. In Lecture Notes in Artificial Intelligence - ISSN:0302-9743
9. Desassis, N., Renard, D.: Automatic variogram modeling by iterative least squares: univariate and multivariate cases. Math. Geosci. **45**, 453–470 (2013). https://doi.org/10.1007/s11004-012-9434-1
10. Ech, M., Morel, S., Pouteau, B., Yotte, S., Breysse, D.: Laboratory evaluation of pavement macrotexture durability. Revue Européenne de Génie Civil, **11**, 5, 643–662 (2007). http://dx.doi.org/10.1080/17747120.2007.9692949
11. Evangelisti, A., Katicha, S., Izeppi, E., Flintsch, G., D'Apuzzo, M., Nicolosi, V.: Measurement error models (MEMs) regression method to harmonize friction values from different skid testing devices (2016). https://doi.org/10.1002/9781119318583.ch12, pp. 159-173. In Materials and Infrastructures, vol. 1,5A - ISBN:9781119318583

12. Haynes, K., Fearnhead, P., Eckley, I.A.: A computationally efficient nonparametric approach for changepoint detection. Stat. Comput. **27**(5), 1293–1305 (2017). ISSN 1573-1375. https://doi.org/10.1007/s1122201696875, https://doi.org/10.1007/s11222-016-9687-5
13. James, N.A., Matteson, D.S.: ECP: An R package for nonparametric multiple change point analysis of multivariate data. J. Stat. Softw. **62**(7) (2014). https://www.jstatsoft.org/
14. Killick, R., Fearnhead, P., Eckley, I.A.: Optimal detection of changepoints with a linear computational cost. J. Am. Stat. Assoc. **107**(500), 1590–1598 (2012)
15. Lebas, M., Peybernard, J., Carta, V.: Méthod de traitement des enregistrements de mesure de densità en continu. Bulletin Liason Laboratoire des Ponts et Chaussées n. 114, Juillet-août (1981)
16. Losa, M., Leandri, P.: The reliability of tests and data processing procedures for pavement macrotexture evaluation. Int. J. Pavement Eng. **12**(1), 59–73, Taylor and Francis (2011). https://doi.org/10.1080/10298436.2010.501866
17. MINES ParisTech/ARMINES: RGeostats: The Geostatistical R Package. Version: 12.0.0. (2020). Free download from: http://cg.ensmp.fr/rgeostats
18. Olea, Ricardo A.: A six-step practical approach to semivariogram modelling. Stoch. Environ. Res. Risk Assess **20**, 307–318 (2006). https://doi.org/10.1007/s00477-005-0026-1
19. PIARC: Optimization of Pavement Surface Characteristics, PIARC Technical Committee on Surface Characteristics, Report to the XVIIIth World Road Congress, Brussels, Belgium (1987)
20. R Core Team: R: A language and environment for statistical computing. R Foundation for Statistical Computing, Vienna, Austria (2020). https://www.R-project.org/
21. Saroli, M., Albano, M., Modoni, G., Moro, M., Milana, G., Spacagna, R.L., Falcucci, E., Gori, S., Scarascia Mugnozza, G.: Insights into bedrock paleomorphology and linear dynamic soil properties of the Cassino intermontane basin (Central Italy). In Engineering Geology - Volume 264 (2020) 105333 - https://doi.org/10.1016/j.enggeo.2019.105333
22. Spacagna, R.L., Modoni, G.: Gis-based study of land subsidence in the city of Bologna. In: Mechatronics for Cultural Heritage and Civil Engineering, pp. 235–256 (2018). https://doi.org/10.1007/978-3-319-68646-2_10
23. Thomas, F.: Statistical approach to road segmentation. ASCE J. Transp. Eng. **129**(3), 300–308 (2003). https://doi.org/10.1061/(ASCE)0733-947x(2003)129:3(300)

Directional Supervised Learning Through Depth Functions: An Application to ECG Waves Analysis

Houyem Demni

Abstract The present work investigates arrhythmias which can be detected from Electrocardiography (ECG) waves. Detecting cardiac arrhythmia helps indeed to prevent sudden and untimely deaths. Here, directional depth-based classifiers are employed to predict the presence or absence of cardiac arrhythmia. A comparison of their performance with respect to the directional Bayes rule is also provided.

Keywords Distance-based depth · Directional variables · Supervised classification

1 Introduction and Motivations

Over many decades, linearization was used to explore spherical data by trying to circumvent their non-linear nature. Then, Fisher [5] showed that linear approximations hamper studying some specific phenomena such as the remanent magnetism in sedimentary rocks. Thereafter, several studies have been dedicated to analyze directional data in an appropriate way due to their distinctive properties (e.g. [8, 11]).

The use of directional statistical methods has been motivated by interesting applications in many fields such as astronomy, bioinformatics, neurology, genetics, aeronautics, medicine, and machine learning. Here, we focus on the application of directional supervised learning techniques to Electrocardiography (ECG) wave analysis. The aim is to find a function that assigns new patients to either the class of healthy or ill people, based on values obtained from their ECG waves. To this end, the predictive variables in our problem are not treated as linear continuous variables anymore but as directional variables measured in angles.

Within the context of directional supervised classification, new depth-based classifiers have been quite recently introduced: the max-depth classifier [14], the DD-

H. Demni (✉)
University of Cassino and Southern Lazio, Cassino, Italy
e-mail: houyem.demni@unicas.it

Laboratoire ARBRE, Institut Supérieur de Gestion de Tunis, Université de Tunis, Tunis, Tunisia

S. Balzano et al. (eds.), *Statistical Learning and Modeling in Data Analysis*,
Studies in Classification, Data Analysis, and Knowledge Organization,
https://doi.org/10.1007/978-3-030-69944-4_9

classifier [13], and the depth distribution classifier [2]. For these classifiers, both optimality properties and simulation results are available.

Vencalek et al. [17] derived the conditions under which some of these classifiers are optimal in the Bayes sense. For instance, they found that the max-depth and the depth distribution classifiers are optimal if the underlying distributions are rotationally symmetric, unimodal, differ only in location, and have equal prior probabilities.

Robustness properties of the max-depth, depth versus depth and depth distribution classifiers were investigated under different contamination schemes in [3]. It came out that the DD-classifier performs better or equivalent to the empirical Bayes while it outperforms the max-depth and the depth distribution classifiers in the presence of noise.

What is still lacking is to evaluate how these depth-based directional classifiers perform on real data. This short note has thus the goal of starting fulfilling this gap. With that aim, this work analyzes the performance of the max-depth, the depth versus depth, and the depth distribution classifiers on a real data set which is well known in the supervised learning literature. It refers to some arrhythmia data used to discriminate between healthy and ill people. In our study, we focus on the directional predictors which come from ECG waves. The performance of such classifiers is also compared with the performance of the directional Bayes classifier under the hypothesis of a von Mises-Fisher distribution.

The work is organized as follows. Section 2 presents the arrhythmia data set, a description of the directional variables, and the overall aim of the analysis. In Sect. 3, we briefly present the mentioned depth-based classifiers for directional data. Section 4 reports results on the performance of the depth-based classifiers when applied to the ECG waves problem. In Sect. 5, some final remarks are offered to the reader.

2 The Arrhythmia Data Set

Arrhythmia refers to irregular heartbeats, and it can be evaluated by looking at the electrical activity of the heart, recorded through Electrocardiogram (ECG) waves. Analyzing ECG waves can provide insights into heart health issues. These waves can be in turn be treated as angular variables.

The arrhythmia data [16] is one of the data sets available within the UCI Machine Learning Repository [6]. It reports the presence of different types of cardiac-arrhythmia from ECG as well as its absence. The original data set contains 452 patient records described by 279 predictive variables (measurements, patient data, and ECG recordings) and 16 classes: the first refers to normal ECG (healthy patients) while classes 2–15 correspond to different types of arrhythmia and class 16 refers to the unclassified patients.

2.1 Previous Studies

Among the many, the studies which exploited arrhythmia data while adopting directional data techniques are considered here. All of them dealt with the waves as angular variables.

First, Lopez-Cruz et al. [10] proposed an extension to directional data of the naive Bayes classifier and the selective naive Bayes for von Mises and von Mises-Fisher distributions. They showed their superiority with respect to other versions of naive Bayes.

Then, Fernandes and Cardoso [4] introduced a discriminative binary classifier for mixed data (linear and angular), and they showed that their method is competitive to traditional classifiers. More recently, Pernes et al. [15] proposed several versions of directional support vector machines that support both angular and linear predictors and compared them to several directional classifiers.

The best average misclassification rate for the arrhythmia data is 0.209, and it was obtained in [15] by a directional logistic regression model. However, we note that the performances reported within these three papers are not directly comparable, given that different simulation settings have been adopted within each of them. Furthermore, such performances are not comparable with our case study results as well, where the focus is on the discriminant power of the directional variables on their own.

2.2 Scope of the Analysis and Variable Description

In line with the previously mentioned studies [4, 10, 15], unclassified samples were removed and the study goal was transformed into a binary classification problem (normal vs. arrhythmia).

As predictors, the four angular variables characterizing ECG waves are considered. That is, the aim of our study is to discriminate between healthy and non-healthy patients with arrhythmia based on the values obtained from their ECG waves. Table 1 summarizes the number of directional variables, the number of classes, and the number of observations per class of the evaluated data set.

Table 1 Summary of the main characteristics of the data used in this work, including the number of directional (dir.) features and the number of observations (obs.) per class (class 1—normal vs. class 2—arrhythmia)

Number of dir. variables	Classes	Number of obs.	Number of obs. per class
4	2	430	Class 1, normal: 245 Class 2, arrhythmia: 185

The angular variables characterize the vector angles from the front plane of four ECG waves, and they are measured in degrees in the original data set. The P-wave reflects the atrial depolarization, the QRS-wave represents the depolarization of the ventricles, the T-wave describes the rapid re-polarization of contractile cells while the QRST-wave corresponds to the global ventricular re-polarization.

By looking at the rose diagram of each observed distribution separately for the groups of healthy and ill people, we saw they are unimodal. Hence, their distribution can be properly investigated by means of circular box-plots [1], which are here represented in Fig. 1.

We note that the distribution of the QRS-wave angles span over more than half a circle, while all the others have angles in $(0, \pi)$. As a consequence, they can be mapped into a 4-dimensional hyper-sphere embedded in a $5D$ space. Directional supervised learning procedures act directly on such a hyper-sphere.

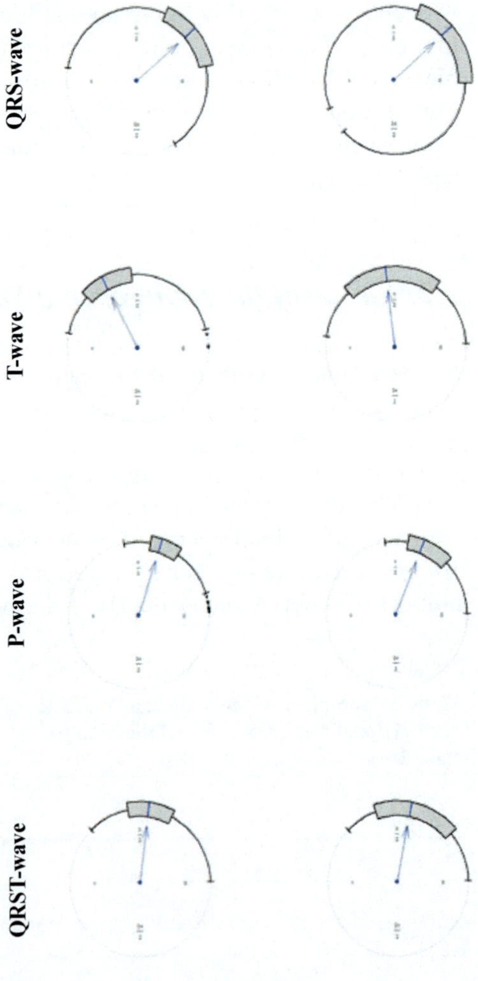

Fig. 1 Circular box-plots of the angular variables exploited in this study. By column: healthy patients (left) and patients with arrhythmia (right). By row: QRS-wave, T-wave, P-wave, and QRST-wave

Some of the observed marginal distributions are substantially symmetric, others are clearly asymmetrically distributed (e.g. the P-wave angles and, to a certain extent, the QRST waves). Looking at differences between the two groups, the T-waves seem to have the higher marginal ability to discriminate.

3 Directional Depth-Based Supervised Learning Techniques

In this section, the three main directional depth-based supervised classification methods are briefly reviewed: the max-depth classifier, the depth versus depth classifier (DD-classifier), and the depth distribution classifier.

Considering K empirical distributions $\hat{H}_i, i = 1, .., K$, the directional max-depth classifier is given by

$$class_{max}(x) := \text{argmax}_i D(x; \hat{H}_i),$$

where $x \in S^{(q-1)}$ is a new observation to be classified, and $D(x, \hat{H}_i)$, $i = 1, .., K$ is the empirical depth of x with respect to the directional empirical distributions $\hat{H}_1, .., \hat{H}_K$, respectively.

The directional DD-classifier is a generalization of the max-depth classifier and it is given by

$$class_{DD}(x) := \underset{i}{\text{argmax}} r(D(x; \hat{H}_i)), \tag{1}$$

where $r(.)$ is a real increasing function which has the aim of well separated points in the depth versus depth space (DD-plot). Different choices have been considered for $r(.)$. Li et.al. [9] suggested considering a polynomial discriminating function, whose degree has to be estimated, while Mosler and Mozharovskyi [12] adopted a k-NN decision rule.

The directional depth distribution classifier is given by

$$class_{dd}(x) := \text{argmax}_i F_D(x, \hat{H}_i),$$

with

$$F_D(x, \hat{H}_i) := P(D(X, \hat{H}_i) \leq D(x, \hat{H}_i)),$$

where $D(x, \hat{H}_i), i = 1, .., K$ is the empirical depth of x with respect to the empirical distributions $\hat{H}_1, .., \hat{H}_K$, respectively, and hence $F_D(., \hat{H}_i)$ is the cdf of the depth function under $\hat{H}_i$.

For each classifier, a depth function must be adopted. Here, distance-based depth functions are considered. They are defined as follows [14]:

- The cosine depth: $D_{cos}(x, H) = 2 - E_H[(1 - x'X)]$;
- The arc-distance depth: $D_{arc}(x, H) = \pi - E_H[\arccos(x'X)]$;
- The chord depth: $D_{chord}(x, H) = 2 - E_H[\sqrt{2(1 - x'X)}]$.

Here, $x \in S^{(q-1)}$ is a point whose depth is evaluated with respect to the directional distribution H, $E[.]$ is the expected value, and X is a random variable from H. The empirical depth is obtained by replacing H with $\hat{H}$ for each depth function.

Finally, for the sake of completeness, we recall how the empirical Bayes classifier is defined. We have

$$class_{Bayes}(x) := \text{argmax}_i \hat{h}_i(x) p_i$$

where p_i is the prior probability corresponding to the distribution H_i, $i = 1, .., K$, and $\hat{h}_i()$ is the estimated assumed density for the ith group. In directional supervised learning, the Bayes classifiers has been used with the $h_i()$'s being von Mises-Fisher densities with different location and concentration parameters [10].

4 Performance of Depth-Based Classifiers on ECG Waves

As discussed, the aim of this study is to evaluate the performance of depth-based classifiers on a set of real data arising from an ECG analysis. With that goal, the angular variables were transformed to their Euclidean coordinates (units vectors) and a simulation study was performed. In line with the existing literature [4, 10, 15], a threefold stratified cross-validation method where the percentage of samples for each class is preserved was considered. The experiment was repeated 100 times.

Ten different possible solutions were evaluated and compared. Each of the three mentioned classifiers was combined with three different directional depth functions (cosine, chord, and arc-distance), and all of them were compared against the empirical Bayes classifier under the von Mises-Fisher assumption.

For the $r(.)$ function in Eq. 1, the k-Nearest Neighborhood (k-NN) discriminant rule has been adopted in line with [3, 14], with the tuning parameter k chosen by cross validation. The performance of the classifiers was evaluated by means of the misclassification rate which is the number of misclassified observations over the sample size in each replicated sample, and by the macro F_1-score which is the unweighted mean value of the individual F_1-scores of each class.

The distribution of the misclassification rates obtained by the max-depth, depth distribution, DD- and Bayes classifiers when associated with different distance-based depth functions are here provided through box-plots (Fig. 2) and summarized through the average misclassification rates (Table 2). The macro F_1-scores of the directional classifiers are also given in Table 2.

Although the two classes are imbalanced, the average macro F_1-score and the average accuracy of the classifiers are consistent with each other: the best classifier in terms of average accuracy is the best classifier in terms of average macro F_1-score too (Table 2).

The DD-classifier achieves the best overall performance in terms of average misclassification rate (Fig. 2, most right graph-box). Furthermore, it performs better than the Bayes rule independently from the choice of the depth function. The depth dis-

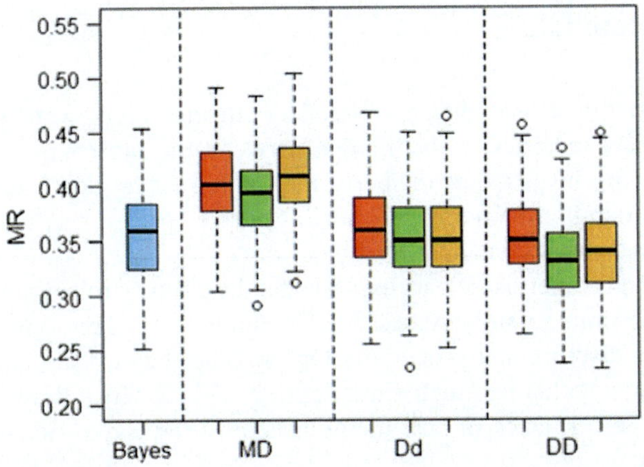

Fig. 2 Box-plots of misclassification rates (MR) of the Bayes, max-depth (MD), depth distribution (Dd), and DD-classifiers (DD) when associated with the cosine, chord, and arc-distance depth functions. In each graph-box (excluding the Bayes), the most left box-plot refers to the cosine depth, the middle one to the chord depth, and the most right to the arc-distance depth. The best performance is achieved by the DD-classifier associated with the chord depth

Table 2 Average misclassification rate (AMR) and average macro F_1-score of the Bayes, max-depth, depth distribution, and DD-classifiers when associated with the cosine, chord, and arc-distance depth functions. Best achieved results are highlighted in bold

Classifier		Average misclassification rate (AMR)	Average macro F_1-score
Empirical Bayes		0.36	0.63
Max-depth	Cosine	0.40	0.60
	Chord	0.39	0.60
	Arc-distance	0.42	0.52
Depth distribution	Cosine	0.36	0.63
	Chord	0.35	0.64
	Arc-distance	0.35	0.63
DD-plot with k-NN	Cosine	0.35	0.64
	Chord	**0.33**	**0.67**
	Arc-distance	0.34	0.66

tribution and the DD-classifiers perform equivalent to the empirical Bayes classifier (if not slightly better). The worst performance is given by the max-depth classifier.

In general, the choice of the depth function seems not to be particularly influential on the performance of the three classifiers, although some small differences arise. In addition, by looking at the confusion matrix of the classifiers, it appears that it is in general more difficult to classify patients with arrhythmia. The higher proportion of misclassified observations arises indeed from class 2 (observations are wrongly assigned to class 1 while they are coming from class 2).

5 Final Remarks

In this work, directional distance-based depth classifiers were applied to some arrhythmia data in order to distinguish between the presence or absence of cardiac diseases. We investigated the performance of the max-depth, depth distribution, depth versus depth classifiers, and the Bayes rule. Angular variables arising from ECG recordings were considered.

In directional supervised learning, the standard Bayes rule assumes data in each group coming from a von Mises-Fisher distribution. If so, the Bayes rule yields the best available discriminant procedure. On the other hand, real data not necessarily fulfill such or any other parametric assumption. This is why it is always of interest to compare the performance of new methods against the Bayes rule on specific fields of application.

On the considered data, we had that the DD-classifier largely outperforms max-depth and it performs better than the depth distribution and the empirical Bayes classifier. On the other hand, the performance of the depth distribution classifier is equivalent to the Bayes rule, and the max-depth classifier definitely provides the worst behavior over all the considered methods.

As further research, we see the necessity of developing new depth-based methods which combine both linear and directional variables to fully exploit the information available within the data set. It would be also of interest to test the discussed directional supervised learning methods on other real data applications within this field.

Acknowledgements The author would like to thank the Editors and the Referees for their helpful comments and suggestions which improved the presentation of this work.

References

1. Buttarazzi, D., Pandolfo, G., Porzio, G.C.: A boxplot for circular data. Biometrics **74**(4), 1492–1501 (2018)
2. Demni, H., Messaoud, A., Porzio, G.C.: The cosine depth distribution classifier for directional data. In: Bauer, N., Ickstadt, K., Lübke, K., Szepannek, G., Trautmann, H., Vichi, M. (eds.), Applications in Statistical Computing. Studies in Classification, Data Analysis, and Knowledge Organization, Chapter 4, pp. 49–60. Springer, Cham (2019). https://doi.org/10.1007/978-3-030-25147-5-4
3. Demni, H., Messaoud, A., Porzio, G.C.: Distance-based directional depth classifiers: a robustness study. Commun. Stat. Simul. Comput., in press (2021)
4. Fernandes, K., Cardoso, J.S.: Discriminative directional classifiers. Neurocomputing **207**, 141–149 (2016)
5. Fisher, R.A.: Dispersion on a sphere. Proc. R. Soc. Lond. Ser. A. Math. Phys. Sci. **217**(1130), 295–305 (1953)
6. Frank, A., Asuncion, A.: UCI Machine Learning Repository (2010). http://archive.ics.uci.edu/ml
7. Frénay, B., Verleysen, M.: Classification in the presence of label noise: a survey. IEEE Trans. Neural Netw. Learn. Syst. **25**(5), 845–869 (2013)

8. Jupp, P.E., Mardia, K.V.: A unified view of the theory of directional statistics, 1975–1988. International Statistical Review/Revue Internationale de Statistique, 261–294 (1989)
9. Li, J., Cuesta-Albertos, J.A., Liu, R.Y.: DD-classifier: nonparametric classification procedure based on DD-plot. J. Am. Stat. Assoc. **107**(498), 737–753 (2012)
10. López-Cruz, P.L., Bielza, C., Larrañaga, P.: Directional naive Bayes classifiers. Pattern Anal. Appl. **18**(2), 225–246 (2015)
11. Mardia, K.V.: Statistics of directional data. J. R. Stat. Soc.: Ser. B (Methodol.) **37**(3), 349–371 (1975)
12. Mosler, K., Mozharovskyi, P.: Fast DD-classification of functional data. Stat. Papers **58**(4), 1055–1089 (2017)
13. Pandolfo, G., D'Ambrosio, A., Porzio, G.C.: A note on depth-based classification of circular data. Electron. J. Appl. Stat. Anal. **11**(2), 447–462 (2018)
14. Pandolfo, G., Paindaveine, D., Porzio, G.C.: Distance-based depths for directional data. Canadian J. Stat. **46**(4), 593–609 (2018)
15. Pernes, D., Fernandes, K., Cardoso, J.S.: Directional support vector machines. Appl. Sci. **9**(4), 725 (2019)
16. UCI ML Repository. Arrhythmia dataset. https://archive.ics.uci.edu/ml/datasets/Arrhythmia (2020). Accessed 3 June 2020
17. Vencalek, O., Demni, H., Messaoud, A., Porzio, G.C.: On the optimality of the max-depth and max-rank classifiers for spherical data. Appl. Math. **65**(3), 331–342 (2020)

Penalized Versus Constrained Approaches for Clusterwise Linear Regression Modeling

Roberto Di Mari, Stefano Antonio Gattone, and Roberto Rocci

Abstract Several approaches exist to avoid singular and spurious solutions in maximum likelihood (ML) estimation of clusterwise linear regression models. We propose to solve the degeneracy problem by using a penalized approach: this is done by adding a penalty term to the log-likelihood function which increasingly penalizes smaller values of the scale parameters, and the tuning of the penalty term is done based on the data. Another traditional solution to degeneracy consists in imposing constraints on the variances of the regression error terms (constrained approach). We will compare the penalized approach to the constrained approach in a simulation study, providing practical guidelines on which approach to use under different circumstances.

Keywords Clusterwise linear regression · Penalized likelihood · Scale constraints

1 Introduction

Let $y_1, \ldots, y_n$ be a sample of independent observations drawn from the response random variable Y_i, each observed alongside with a vector of J explanatory variables $\mathbf{x}_1, \ldots, \mathbf{x}_n$. Let us assume $Y_i|\mathbf{x}_i$ to be distributed as a finite mixture of linear regression models, that is

R. Di Mari (✉)
Department of Economics and Business, University of Catania, Catania, Italy
e-mail: roberto.dimari@unict.it

S. A. Gattone
Department of Philosophical and Social Sciences, Economics and Quantitative Methods, University G. d'Annunzio, Chieti-Pescara, Italy
e-mail: gattone@unich.it

R. Rocci
Department of Statistical Sciences, University of Rome La Sapienza, Rome, Italy
e-mail: roberto.rocci@uniroma1.it

S. Balzano et al. (eds.), *Statistical Learning and Modeling in Data Analysis*, Studies in Classification, Data Analysis, and Knowledge Organization,
https://doi.org/10.1007/978-3-030-69944-4_10

$$f(y_i|\mathbf{x}_i;\boldsymbol{\psi}) = \sum_{g=1}^{G} p_g \phi_g(y_i|\mathbf{x}_i, \sigma_g^2, \boldsymbol{\beta}_g) = \sum_{g=1}^{G} p_g \frac{1}{\sqrt{2\pi\sigma_g^2}} \exp\left[-\frac{(y_i - \mathbf{x}_i'\boldsymbol{\beta}_g)^2}{2\sigma_g^2}\right], \tag{1}$$

where G is the number of clusters and p_g, $\boldsymbol{\beta}_g$, and σ_g^2 are the mixing proportion, the vector of $J+1$ regression coefficients that includes an intercept, and the variance term for the gth cluster. The set of all model parameters is given by $\boldsymbol{\psi} = \{(p_1,\ldots,p_G;\boldsymbol{\beta}_1,\ldots,\boldsymbol{\beta}_G;\sigma_1^2,\ldots,\sigma_G^2) \in \mathbb{R}^{(G-1)+(J+1)G+G} : p_1+\cdots+p_G = 1, p_g > 0, \sigma_g^2 > 0, \text{ for } g = 1,\ldots,G\}$.

The likelihood function can be specified as

$$\mathscr{L}(\boldsymbol{\psi}) = \prod_{i=1}^{n}\left\{\sum_{g=1}^{G} p_g \frac{1}{\sqrt{2\pi\sigma_g{}^2}} \exp\left[-\frac{(y_i - \mathbf{x}_i'\boldsymbol{\beta}_g)^2}{2\sigma_g^2}\right]\right\}, \tag{2}$$

which we maximize to estimate $\boldsymbol{\psi}$ either by means of direct maximization or with the perhaps more popular EM algorithm [5]. However, there is a well-known complication in ML estimation of this class of models: the likelihood function of mixtures of (conditional) normals with cluster-specific variances is unbounded [4, 11].

A traditional solution to the problem of unboundedness is based on the seminal work of [7] which, for univariate mixtures of normals, suggested imposing a lower bound to the ratios of the scale parameters in the maximization step. The method is equivariant under linear affine transformations of the data. That is, if the data are linearly transformed, the estimated posterior probabilities do not change and the clustering remains unaltered. Recently, in the multivariate case, [12] incorporated constraints on the eigenvalues of the component covariances matrices of Gaussian mixtures that are tuned on the data based on a cross-validation strategy. These constraints are built upon [9]'s reformulation and are an equivariant sufficient condition for Hathaway's constraints. Estimation is done in a familiar ML environment [10], with a data-driven selection of the scale balance. Di Mari et al. [6] adapted [12]'s method to clusterwise linear regression, further investigating its properties.

Another possible approach for handling unboundedness is to modify the log-likelihood function by adding a penalty term, in which smaller values of the scale parameters are increasingly penalized. Representative examples can be found in [1–3].

In this work, we review the constrained approach of [6] and develop a data-driven equivariant penalized approach for ML estimation. In Sect. 2, we sketch the bulk of the methodologies; in Sect. 3 we report the results from the simulation study and then draw some conclusions (Sect. 4).

2 The Methodology

2.1 *The Constrained Approach*

Di Mari et al. [6] proposed relative constraints on the group conditional variances σ_g^2 of the kind

$$\sqrt{c} \leq \frac{\sigma_g^2}{\bar{\sigma}^2} \leq \frac{1}{\sqrt{c}}, \tag{3}$$

or equivalently

$$\bar{\sigma}^2\sqrt{c} \leq \sigma_g^2 \leq \bar{\sigma}^2\frac{1}{\sqrt{c}}. \tag{4}$$

The above constraints are equivariant and have the effect of shrinking the variances to a suitably chosen $\bar{\boldsymbol{\sigma}}^2$, the *target* variance term, and the level of shrinkage is given by the value of c. These constraints are easily implementable within the EM algorithm [9, 10], which is fully available in closed form, and the selection of c is based on the data.

2.2 *The Penalized Approach*

An alternative to the constrained estimator is the penalized approach, in which a penalty $s_n(\sigma_1^2, \ldots, \sigma_G^2)$ is put on the component variances and it is added to the log-likelihood. Under certain conditions on the penalty function, the penalized estimator is know to be consistent [1]. A function s_n that satisfies these conditions is

$$s_n(\sigma_1^2, \ldots, \sigma_G^2) = -\lambda \sum_{g=1}^{G} \left(\frac{\bar{\sigma}^2}{\sigma_g^2} + \log(\sigma_g^2) \right), \tag{5}$$

where $\bar{\sigma}^2$, the *target* variance, can be seen as our *prior* information on the scale structure and λ is the penalizing constant that is selected based on the data. Thus, the penalized log-likelihood can be written as

$$p\ell(\boldsymbol{\psi}) = \ell(\boldsymbol{\psi}) + s_n(\sigma_1^2, \ldots, \sigma_G^2) \tag{6}$$

and the set of unknown parameters is found by ML with computation done by means of an EM algorithm that is available in closed form. Besides the constrained approach, the penalized approach is equivariant with respect to linear transformation in the response.

2.3 Selection of the Tuning Parameter

Both approaches require selection of the tuning parameter—c and λ, respectively, for the constrained and penalized estimators. The tuning constants can be pre-specified by the user if any prior knowledge on the scale structure of the cluster is available. If this is not the case, the tuning can be based on the data. We propose two alternative approaches to select the tuning constant that can be used for both constrained and penalized methods.

2.3.1 Cross-Validation

The first tuning approach is based on a cross-validation strategy that looks for a tuning parameter such that the cross-validated likelihood is maximized. For a given c or λ, this is done as follows:

1. Temporary estimates for the model parameters are obtained from the entire sample, and these are used as starting values to initialize the cross-validation procedure.
2. The data set is partitioned into training and test sets.
3. Parameters are estimated on the training set and the contribution to the log-likelihood of the test set is computed.
4. Steps 2–3 are repeated M times and the M contributions to the log-likelihood of the test set are summed for different values of c/λ.

2.3.2 *k*-Deleted Method

The second tuning approach is based on the modification of the k-deleted method [13, 14] that looks for a tuning parameter such that the (modified) k-deleted log-likelihood[1] is maximized.

For a given c or λ, this is done as follows:

1. Temporary estimates for the model parameters are obtained from the entire sample, and these are used as starting values to initialize the procedure.
2. For a given c/λ, the model parameters are estimated.
3. The (modified) k-deleted log-likelihood is computed.
4. Steps 2–3 are repeated for different values of c/λ.

[1]For some estimates of the model parameters, this is computed by taking out the k units with the largest log-likelihood.

3 Simulation Study

A simulation study has been conducted to compare the quality of the parameter estimates and the ability to recover the clusters structure of the constrained and the penalized approaches. Both tuning strategies—cross-validation based and k-deleted method—were considered for the constrained and penalized approaches—respectively *conC*, *conCk*, *penC*, and *penCk*—and the unconstrained estimator with common (homoscedastic) component-scales (hom) and the unconstrained estimator with different (heretoscedastic) component-scales (het) were also included for comparison.

The target measures used for the comparisons were average Mean Squared Errors (MSE) of the regression coefficients (averaged across regressors and groups) and the adjusted Rand index [8].

We generated the data from a 3-group clusterwise linear regression model with 3 regressors and an intercept term. The group mixing weights were set equal to 0.1, 0.3, and 0.6. The regressors were generated from 3 independent standard normal distributions; regression coefficients were randomly generated from Uniform distributions U(−1.5, 1.5), and the group-specific intercepts were set equal to 4, 9, and 16.

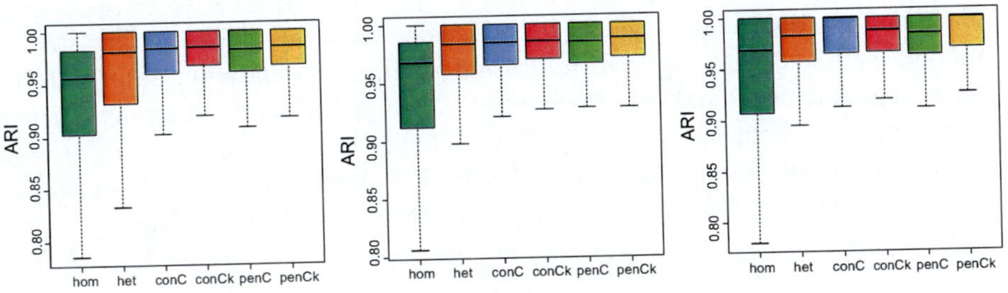

(a) $\sigma^2 = (0.1, 0.8, 0.1)$ (b) $\sigma^2 = (0.2, 0.6, 0.2)$ (c) $0.5, 0.5, 0.5$

Fig. 1 (average) MSE of the regression coefficients for all approaches, for the three scale scenarios and $n = 100$

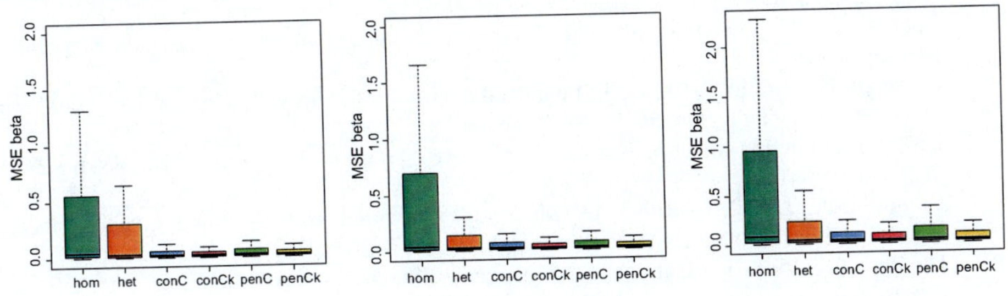

(a) $\sigma^2 = (0.1, 0.8, 0.1)$ (b) $\sigma^2 = (0.2, 0.6, 0.2)$ (c) $0.5, 0.5, 0.5$

Fig. 2 Adjusted Rand Index (ARI) for all approaches, for the three scale scenarios and $n = 100$

We considered 6 crossed simulation conditions of sample size—$n = 100, 200$—and scale scenarios—$\boldsymbol{\sigma}^2 = (0.1, 0.8, 0.1)'$ (*heteroscedasticity*), $\boldsymbol{\sigma}^2 = (0.2, 0.6, 0.2)'$ (*mild heteroscedasticity*), and $\boldsymbol{\sigma}^2 = (0.5, 0.5, 0.5)'$ (*homoscedasticity*)

For each simulation condition, we generated 250 samples and, for each approach, we selected the best solution (highest likelihood) out of 10 random starts. We report only the results for $n = 100$ as those for $n = 200$ were qualitatively the same (Figs. 1 and 2).

We observe that the penalized and constrained approaches overcome their unconstrained rivals (hom and het) both in terms of quality of regression parameter estimates and cluster recovery. It seems that while with a tuning based on the more time-consuming cross-validation strategy conC does slightly better than penC, with the more efficient k-deleted tuning the penalized approach penCk does better than conCk. Overall, penCk delivers the best performance.

4 Concluding Remarks

In this work, we have proposed a new penalized estimator for clusterwise linear regression models in which penalties are put on the component scales. This penalized estimator is equivariant under changes in the scale of the response. We have compared it with the constrained approach of [6] and illustrated two alternative tuning strategies for both methodologies. The constrained and penalized estimators perform uniformly better than unconstrained ones. Whenever the computing time of tuning strategies is not an issue, both approaches serve well the scope of fitting clusterwise linear regression models. For quicker—and perhaps less-refined selection strategies—like the k-deleted method, the penalized approach seems to be preferable.

References

1. Chen, J., Tan, X.: Inference for multivariate normal mixtures. J. Multivariate Anal. **100**(7), 1367–1383 (2009)
2. Chen, J., Tan, X., Zhang, R.: Inference for normal mixtures in mean and variance. Stat. Sinica, 443–465 (2008)
3. Ciuperca, G., Ridolfi, A., Idier, J.: Penalized maximum likelihood estimator for normal mixtures. Scandinavian J. Stat. **30**(1), 45–59 (2003)
4. Day, N.: Estimating the components of a mixture of normal distributions. Biometrika **56**(3), 463–474 (1969)
5. Dempster, A., Laird, N., Rubin, D.: Maximum likelihood from incomplete data via the em algorithm. J. R. Stat. Soc.: Ser. B (Methodol.) **39**(1), 1–22 (1977)
6. Di Mari, R., Rocci, R., Gattone, S.: Clusterwise linear regression modeling with soft scale constraints. Int. J. Approx. Reas. **91**, 160–178 (2017)
7. Hathaway, R.: A constrained formulation of maximum-likelihood estimation for normal mixture distributions. Ann. Stat. **13**(2), 795–800 (1985)
8. Hubert, L., Arabie, P.: Comparing partitions. J. Classif. **2**(1), 193–218 (1985)

9. Ingrassia, S.: A likelihood-based constrained algorithm for multivariate normal mixture models. Stat. Methods Appl. **13**(2), 151–166 (2004)
10. Ingrassia, S., Rocci, R.: Constrained monotone em algorithms for finite mixture of multivariate gaussians. Comput. Stat. Data Anal. **51**(11), 5339–5351 (2007)
11. Kiefer, J., Wolfowitz, J.: Consistency of the maximum likelihood estimator in the presence of infinitely many incidental parameters. Ann. Math. Stat. 887–906 (1956)
12. Rocci, R., Gattone, S., Di Mari, R.: A data driven equivariant approach to constrained gaussian mixture modeling. Advanc. Data Anal. Classif. **12**(2), 235–260 (2018)
13. Seo, B., Kim, D.: Root selection in normal mixture models. Comput. Stat. Data Anal. **56**(8), 2454–2470 (2012)
14. Seo, B., Lindsay, B.G.: A computational strategy for doubly smoothed mle exemplified in the normal mixture model. Comput. Stat. Data Anal. **54**(8), 1930–1941 (2010)

Effect Measures for Group Comparisons in a Two-Component Mixture Model: A Cyber Risk Analysis

Maria Iannario and Claudia Tarantola

Abstract This article deals with ordinal effect measures in the CUP models. They are mixture models for ordinal data with an uncertainty component, where the Uniform distribution is used to model indecision and the standard cumulative model is employed for the analysis of evaluation. We present probability-based measures for comparing clusters on ratings, while adjusting for other explanatory variables, and discuss marginal effects to address the interpretation of the results on the extreme categories of a cyber risk scale.

Keywords Cumulative logit models · CUP models · Extreme categories · Marginal effects · Ordinal superiority measures · Uncertainty

1 Introduction

A widespread recent literature deals with a class of mixture models for rating data that considers the selection of a response category as a combination of the deliberate choice based on the preference of the respondent and an uncertainty component (see [18, 19] and reference therein). In the basic CUB framework [2, 17], the deliberate choice is modelled by a shifted Binomial distribution and the uncertainty part by a discrete Uniform distribution. Various models with different specifications of the distributions of the considered choice and the uncertainty component have been proposed; see, for example, CUB models with varying uncertainty [6], CUB models for a don't know category [11, 15], nonlinear CUB models [14]. Furthermore, to take into account the possible presence of overdispersion the author of [9] introduced CUBE models, where the shifted Binomial distribution has been substituted with a

M. Iannario (✉)
Department of Political Sciences, University of Naples Federico II, Naples, Italy
e-mail: maria.iannario@unina.it

C. Tarantola
Department of Economics and Management, University of Pavia, Pavia, Italy
e-mail: claudia.tarantola@unipv.it

S. Balzano et al. (eds.), *Statistical Learning and Modeling in Data Analysis*,
Studies in Classification, Data Analysis, and Knowledge Organization,
https://doi.org/10.1007/978-3-030-69944-4_11

shifted Beta Binomial random variable. Finally, CUB models with *shelter* have been presented by the author of [8] to take into account a possible inflation in one of the categories. An overview on the modelling approaches has been presented in [10, 20] whereas a comparison with traditional cumulative models for rating data analysis [16] is in [19].

Starting with the basic cumulative model approach confounded with the attention on the uncertainty, detected when a subject selects a score on a rating question, the authors of [21] introduce the alternative CUP models. They are a *C*ombination of two components referred to the individual indecision (*U*ncertainty), expressed on the selection or motivated by the context, and a deliberate choice of a response category determined by the *P*reference of the respondent. A recent development replaces the Uniform with a Beta Binomial random variable to process the uncertainty [22]. The proposal deals with a more flexible distribution which allows distinguishing between a tendency to middle categories and a tendency to extreme categories.

In this contribution, we focus on CUP models; for this mixture, as a consequence of the nonlinearity, model parameters are not as simple to interpret like slopes and correlations for ordinary linear regression. Thus, effect measures based on marginal effects are discussed (see [13] for a comprehensive analysis). The paper surveys simpler ways to interpret the effects of the explanatory variables simplifying the interpretation of the models, describing and visualizing average marginal effects. Furthermore, the article considers simple ordinal effect summaries for model-based comparisons of groups on ratings, while adjusting for other explanatory variables by following the idea by the authors of [1] for standard models. Section 2 is devoted to the introduction of the model, discussion of marginal effects and presentation of measures for ordinal models. Section 3 shows the interpretative usefulness of the CUP model by investigating marginal effects and group comparison measures in a cyber risk analysis. Section 4 concludes with some final remarks and possible future extensions.

2 CUP Models

In a CUP model, the probability distribution of the ordinal response variable R_i $(i = 1, 2, \dots, n)$, describing the rating assigned by respondent i, is given by

$$P(R_i = r|\boldsymbol{x}_i) = \pi_i P_M(Y_i = r|\boldsymbol{x}_i) + (1 - \pi_i)P(U_i = r), \quad r = 1, 2, \dots, m. \tag{1}$$

The *P*reference part $P_M(Y_i = r|\boldsymbol{x}_i)$ is obtained via a cumulative link model on an appropriate row vector of covariates $\boldsymbol{x}_i = \left(x_{i1}, \dots, x_{ij}, \dots x_{ip}\right)$. More precisely,

$$P_M(Y_i \le r|\boldsymbol{x}_i) = F(\alpha_r - \boldsymbol{x}_i\boldsymbol{\gamma}) \quad i = 1, 2 \dots, n; \quad r = 1, 2 \dots, m-1,$$

where $F(\cdot)$ is the cumulative distribution function, α_r is the threshold of the latent scale Y^* of Y (see [16]), and $\boldsymbol{\gamma}$ is a parameter vector of dimension p.

The uncertainty parameter π_i may also depend on a row vector of covariates $\mathbf{w}_i = (w_{i1}, \ldots, w_{ij}, \ldots w_{iq})$, which may have a non-empty intersection with $\mathbf{x}_i$. A logit link is usually applied to model the effect of covariates on the uncertainty component, $\pi_i = \pi_i(\boldsymbol{\beta}) = 1/(1 + e^{-\mathbf{w}_i\boldsymbol{\beta}})$, where $\boldsymbol{\beta}$ is a parameter vector of dimension q. Finally, as mentioned, the second component of the mixture $P(U_i = r)$ follows a discrete Uniform distribution. For given data $(r_i, \mathbf{x}_i)$, the likelihood contribution of observation i is given by

$$\ell_i(r_i; \boldsymbol{\theta}) = \log\left[\pi_i P_M(Y_i = r|\mathbf{x}_i) + (1 - \pi_i)P(U_i = r)\right] \qquad r = 1, 2, \ldots, m, \tag{2}$$

where $\boldsymbol{\theta} = (\boldsymbol{\alpha}, \boldsymbol{\gamma}, \boldsymbol{\beta})'$ collects all parameters of the ordinal model used in the mixture components. The log-likelihood is $\ell(\boldsymbol{\theta}) = \sum_{i=1}^{n} \ell_i(r_i; \boldsymbol{\theta})$.

A way to obtain stable estimates is to consider the mixture as a problem with incomplete data and use the EM algorithm [3] (see Appendix of [21] for details).

2.1 *Marginal Effect Measures for Covariates in* CUP *Models*

One natural way to interpret the effect of one explanatory variable is to consider the corresponding marginal effects (MEs). A ME shows how a variation in one variable affects the outcome distribution, holding all the other variables constant. MEs are computed differently for continuous and categorical covariates. We refer to [7] for a discussion of the interpretation of MEs in ordered response models. As an exemplification, we report the ME on $P(R_i = 1)$ of an explanatory variable x_{ij} involved only in the preference part of the model. If x_{ij} is continuous, the ME on $P(R_i = 1)$ is given by the partial derivative of $P(R_i = 1)$ with respect to x_{ij}

$$ME_{\{R_i=1\}} = \frac{\partial P(R_i = 1|\mathbf{x}^*_{i\backslash j})}{\partial x_{ij}} = -\pi_i \gamma_j f(\alpha_1 - \mathbf{x}_i\boldsymbol{\gamma}).$$

In the previous equation, $f(\cdot)$ indicates the density function corresponding to the examined cumulative model and $\mathbf{x}^*_{i\backslash j}$ indicates the values assumed by the other explanatory variables. If x_{ij} is a categorical variable, we need to calculate the discrete change. The discrete change for a dichotomous variable x_{ij} is given by

$$ME_{\{R_i=1\}} = \pi_i \times \left[P(R_i = 1|\mathbf{x} = (1, \mathbf{x}^*_{i\backslash j})) - P(R_i = 1|\mathbf{x} = (0, \mathbf{x}^*_{i\backslash j}))\right].$$

If the number of possible values is greater than two, the discrete change is computed as the difference in the predicted probabilities for cases in one category relative to the reference level. For more details on MEs for CUP models, see [13].

2.2 Ordinal Superiority Measures in CUP Models

We discuss ordinal superiority measures for group comparisons in CUP models following the approach presented in [1]. Let us consider a dichotomous covariate d identifying two different groups g_0 and g_1 where $d = 0$ for g_0 and $d = 1$ for g_1. For a specific value $\boldsymbol{x}^*_{\setminus d}$ of the remaining covariates, we define

$$\Delta(\boldsymbol{x}^*_{\setminus d}) = P(R_{g_0} > R_{g_1}) = \sum_{l>k} \hat{\pi}_{0l}(\boldsymbol{x}^*_{\setminus d})\hat{\pi}_{1k}(\boldsymbol{x}^*_{\setminus d}) - \sum_{k>l} \hat{\pi}_{0l}(\boldsymbol{x}^*_{\setminus d})\hat{\pi}_{1k}(\boldsymbol{x}^*_{\setminus d}),$$

where $\hat{\pi}_{0r}(\boldsymbol{x}^*_{\setminus d}) = \hat{P}(R = r; d = 0, \boldsymbol{x}^*_{\setminus d})$ is the fitted value obtained from the examined model for g_0; $\hat{\pi}_{1r}(\boldsymbol{x}^*_{\setminus d})$ is obtained in a similar way for g_1. A value of $\Delta(\boldsymbol{x}^*_{\setminus d})$ greater than zero indicates that it is more likely to obtain a higher rating in g_0 than in g_1. An alternative measure that has null value equal to 0.5 is given by

$$\gamma(\boldsymbol{x}^*_{\setminus d}) = P(R_{g_0} > R_{g_1}) + \frac{1}{2} P(R_{g_0} = R_{g_1}) =$$
$$= \sum_{l>k} \hat{\pi}_{0l}(\boldsymbol{x}^*_{\setminus d})\hat{\pi}_{1k}(\boldsymbol{x}^*_{\setminus d}) + \frac{1}{2} \sum_{l} \hat{\pi}_{0l}(\boldsymbol{x}^*_{\setminus d})\hat{\pi}_{1l}(\boldsymbol{x}^*_{\setminus d}).$$

The previous measures are functionally related, in fact $\Delta(\boldsymbol{x}^*_{\setminus d}) = 2\gamma(\boldsymbol{x}^*_{\setminus d}) - 1$.

If all covariates are continuous with the exception of d, as suggested in [1], a summary measure can be obtained as an arithmetic average of the previous quantities calculated for the observed values of $\boldsymbol{x}^*_{\setminus d}$

$$\Delta = \frac{1}{n} \sum_{\boldsymbol{x}^*_{\setminus d}} \Delta(\boldsymbol{x}^*_{\setminus d}); \qquad \gamma^* = \frac{1}{n} \sum_{\boldsymbol{x}^*_{\setminus d}} \gamma(\boldsymbol{x}^*_{\setminus d}). \tag{3}$$

If all covariates are discrete, we suggest using as summary measures the average obtained considering for $\boldsymbol{x}^*_{\setminus d}$ all possible combinations of the values of the covariates ($|C|$)

$$\Delta^* = \frac{1}{|C|} \sum_{\boldsymbol{x}^*_{\setminus d}} \Delta; \qquad \gamma^* = \frac{1}{|C|} \sum_{\boldsymbol{x}^*_{\setminus d}} \gamma.$$

In the case of discrete and continuous covariates, it is possible to extend (3) by using representative values (mean or median, for instance) for the continuous covariates.

3 Example

Cyber risk commonly refers to any risk of financial loss, disruption or damage to the reputation of an organization resulting from the failure of its information technology systems. Nowadays, with the increasing use of technology, cyber security incidents are rapidly multiplying and no business line can consider itself immune from it. Cyber risk is usually associated with cyber attacks, but its domain is broader and it includes risk events that may be caused without any intention to harm. Cyber risk could materialize in a variety of ways, such as

- deliberate and unauthorized breaches of security to gain access to information systems;
- unintentional or accidental breaches of security;
- operational Information Technology (IT) risks due to factors such as poor system integrity.

Poorly managed cyber risks can leave subjects open to a variety of cybercrimes, with consequences ranging from data disruption to economic destitution.

The paper investigates a sample of 1127 statistical units regarding cyber attacks that occurred worldwide in 2018. We work with a sample of data collected by a group of researchers of Clusit and discussed in their Report of the first semester of 2019. Clusit's experts classify the severity of an attack by an ordinal variable (*Severity*) assuming values 1 (critical severity), 2 (high severity) and 3 (medium severity).

As for explanatory covariates, we examine a dichotomized version of the data. A detailed analysis of the full dataset is provided in [4, 5].

Here we analyse some specific tools of several domains: *Malware* selected among different techniques, *Cybercrime* as the candidate attack, *Bank* as target and *European Union* as the country area of reference. Summary statistics of these dichotomous variables are reported in Table 1. Figure 1 shows frequency distribution of *Severity* highlighting that most of the observations are in the last category (*medium*).

Estimated results of CUP models are reported in Table 2. It lists estimated parameters $\hat{\pi}$, $\hat{\gamma}_p$; $p = 1; 2; 3; 4$ and cutpoints $\hat{\alpha}_r$; $r = 1; 2$ with asymptotic standard errors (in parentheses). No covariates are considered for the uncertainty part. AIC index is 1829.028 compared with a standard cumulative model with $AIC = 1841.615$.

Table 1 Summary statistics concerning the examined characteristics of the 1127 statistical units of the survey

Country
EU (83.50%)–Other (16.50%)
Target
Bank (89.62%)–Other (10.38%)
Technique
Malware (60.42%)–Other (39.58%)
Type of attack
Cybercrime (23.95%)–Other (76.05%)

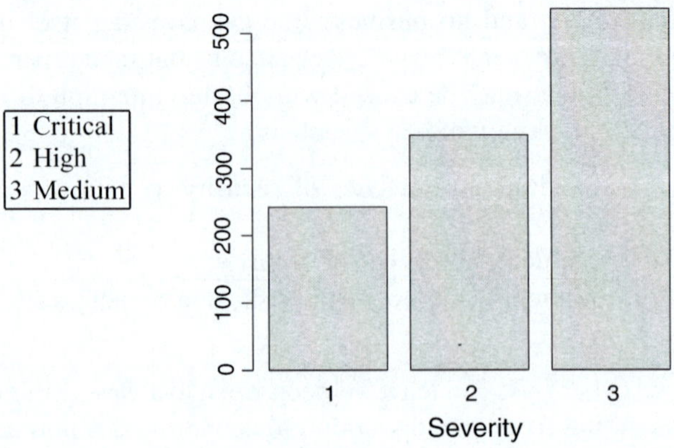

Fig. 1 Frequency distribution of the severity of risk activity

Table 2 Estimated CUP models for Cyber risk analysis

Covariates	Uncertainty parameters	Severity parameters
$\hat{\pi}$	0.790 *(0.029)*	
$\widetilde{EU}$		−0.844*(0.254)*
$\widetilde{Malware}$		−0.772*(0.191)*
$\widetilde{Cybercrime}$		7.001*(1.142)*
$\widetilde{Bank}$		−2.626*(0.384)*
Thresholds	Parameters	
1|2	0.847*(0.215)*	
2|3	5.493*(1.082)*	
$\ell(\boldsymbol{\theta})$	−907.5138	
AIC	1829.028	

Average MEs are in Table 3. Given the decreasing order of the scale, it is possible to observe the reduced effect of the severity of scale when *Cybercrime* is the kind of attack. For the other dichotomous variables, instead, the effect on the margin is the reverse; it means that *EU* area, *Malware* among the technique and *Bank* as target are more likely than the alternative to generate critical severity levels.

These results are confirmed by the summary ordinal superiority measures reported in Table 4. For example, the negative value of Δ^* for *Cybercrime* indicates that there is a higher probability to obtain a medium severity attack in the group g_1. The indexes for the other dichotomous variables, consistently with marginal effects results, indicate that *EU*, *Malware* and *Bank* present higher probability to have the first level (critical) of severity. In same fashion, results of γ^* suggest the ordinal inferiority of g_0 for *EU*, *Malware* and *Bank* and the reverse order for *Cybercrime*.

Table 3 Average Marginal Effect for CUP models—Cyber risk data

$ME.1(Critical)$	Effect	Std.error	z.value	p.value
EU	0.032	0.011	2.988	0.003
Malware	0.029	0.008	3.452	0.001
Cybercrime	−0.264	0.017	−15.753	0.000
Bank	0.099	0.018	5.613	0.000
$ME.3(Medium)$	Effect	Std. error	z.value	p.value
EU	−0.092	0.027	−3.434	0.001
Malware	−0.084	0.020	−4.191	0.000
Cybercrime	0.765	0.102	7.520	0.000
Bank	−0.287	0.031	−9.309	0.000

Table 4 Ordinal superiority measure for CUP models—Cyber risk data

	γ^*	Δ^*
EU	0.541	0.082
Malware	0.538	0.075
Cybercrime	0.140	−0.720
Bank	0.630	0.259

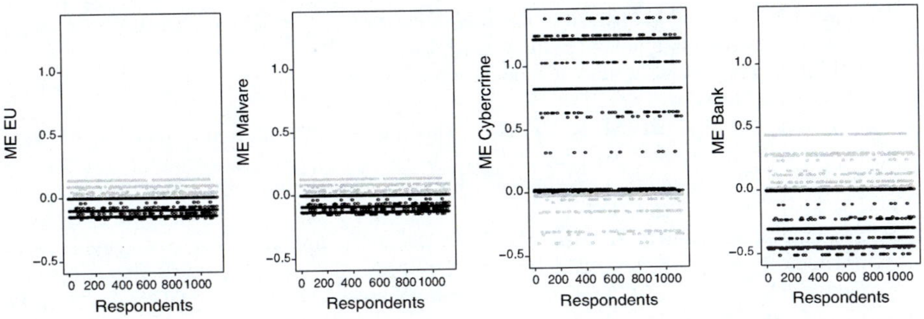

Fig. 2 Group comparisons for individual marginal effects. Light colour is for first marginal effect, dark colour for the last

Finally, Fig. 2 shows the individual marginal effects for the four examined covariates. It underlines a higher variability of *Cybercrime* followed by *Bank* with respect to the other covariates.

The *R* code for the implementation of the results is available in the Supplementary material of [13], whereas the code for the CUP model is available from the first Author under request.

4 Discussion and Extension

The paper discusses effect measures for covariates, especially dichotomous ones, to interpret the cluster contribution in ordinal data models with uncertainty. The measures may be extended also to more general ordinal-response models than those having linear predictors, such as the Betamix and Betabin models (see [22]) or generalized additive models for ordinal responses with uncertainty. The use of the alternative link functions for the Preference part of the models (as probit or complementary log log, for instance) has been discouraged by accounting for robust inference (see [12]). Alternative analyses with the use of different covariates are also planned for the furthered comprehension of cyber risk activities.

Acknowledgements We thank Alan Agresti for his mentoring work, the associate editor and the two referees for their helpful comments. This research has partially been supported by the European Union's Horizon 2020 research and innovation programme "FIN-TECH: A Financial supervision and Technology compliance training programme" under the grant agreement No. 825215 (Topic: ICT-35-2018, Type of action: CSA).

References

1. Agresti, A., Kateri, M.: Ordinal probability effect measures for group comparsons in multinomial cumulative link models. Biometrics **73**, 214–219 (2017)
2. D'Elia, A., Piccolo, D.: A mixture model for preference data analysis. Comput. Stat. Data Anal. **49**, 917–934 (2005)
3. Dempster, A.P., Laird, N.M., Rubin, D.B.: Maximum likelihood from incomplete data via the EM algorithm. J. R. Stat. Soc. Ser. B **39**, 1–38 (1977)
4. Facchinetti, S., Giudici, P., Osmetti. S.: Cyber risk measurement with ordinal data. Stat. Methods Appl. **29**, 173–185 (2020)
5. Facchinetti, S., Osmetti, S., Tarantola, C.: A statistical approach for assessing cyber risk via ordered response models. Under review (2020)
6. Gottard, A., Iannario, M., Piccolo, D.: Varying uncertainty in cub models. Adv. Data Anal. Classif. **10**, 225–244 (2014)
7. Greene, W.H., Hensher, D.A.: Modeling Ordered Choices: A Primer. Cambridge University Press, Cambridge (2010)
8. Iannario, M.: Modelling shelter choices in a class of mixture models for ordinal responses. Stat. Methods Appl. **21**, 1–22 (2012)
9. Iannario, M.: Modelling uncertainty and over dispersion in ordinal data. Commun. Stat. Theory Methods **43**, 771–786 (2013)
10. Iannario, M., Piccolo, D.: A comprehensive framework of regression models for ordinal data. METRON **74**, 233–252 (2016)
11. Iannario, M., Manisera, M., Piccolo, D., Zuccolotto, P.: Ordinal data models for no-opinion responses in attitude survey. Sociol. Methods Res. **49**, 250–276 (2020)
12. Iannario, M., Monti, A.C., Piccolo, D., Ronchetti, E.: Robust inference for ordinal response models. Electron. J. Stat. **11**, 3407–3445 (2017)
13. Iannario, M., Tarantola, C.: How to interpret the effect of covariates on the extreme categories in ordinal data models. Soc. Methods Res. (2021, January)
14. Manisera, M., Zuccolotto, P.: Modelling "Don't know" responses in rating scales. Pattern Recognit. Lett. **45**, 226–234 (2014a)

15. Manisera, M., Zuccolotto, P.: Modeling rating data with Nonlinear cub models. Comput. Stat. Data Anal. **78**, 100–118 (2014b)
16. McCullagh, P.: Regression models for ordinal data (with discussion). J. R. Stat. Soc. Ser. B **42**, 109–142 (1980)
17. Piccolo, D.: On the moments of a mixture of uniform and shifted binomial random variables. Quaderni di Statistica **5**, 85–104 (2003)
18. Piccolo, D., Simone, R.: The class of CUB models: statistical foundations, inferential issues and empirical evidence. Stat. Methods Appl. **1–47**, (2019)
19. Piccolo, D., Simone, R., Iannario, M.: Cumulative and CUB models for rating data: a comparative analysis. Int. Stat. Rev. **87**, 207–236 (2019)
20. Tutz, G.: Ordinal regression: A review and a taxonomy of models. WIREs Comput. Stat. **2021**, e1545. https://doi.org/10.1002/wics.1545
21. Tutz, G., Schneider, M., Iannario, M., Piccolo, D.: Mixture models for ordinal responses to account for uncertainty of choice. Adv. Data Anal. Classif. **11**, 281–305 (2017)
22. Tutz, G., Schneider, M.: Mixture models for ordinal responses with a flexible uncertainty component. J. Appl. Stat. **46**, 1–20 (2018)

A Cramér–von Mises Test of Uniformity on the Hypersphere

Eduardo García-Portugués, Paula Navarro-Esteban, and Juan Antonio Cuesta-Albertos

Abstract Testing uniformity of a sample supported on the hypersphere is one of the first steps when analysing multivariate data for which only the directions (and not the magnitudes) are of interest. In this work, a projection-based Cramér–von Mises test of uniformity on the hypersphere is introduced. This test can be regarded as an extension of the well-known Watson test of circular uniformity to the hypersphere. The null asymptotic distribution of the test statistic is obtained and, via numerical experiments, shown to be tractable and practical. A novel study on the uniformity of the distribution of craters on Venus illustrates the usage of the test.

Keywords Circular data · Craters · Directional data

1 Introduction

Testing uniformity of a sample $\mathbf{X}_1, \ldots, \mathbf{X}_n$ of a random vector $\mathbf{X}$ supported on the hypersphere $\Omega_q := \{\mathbf{x} \in \mathbb{R}^{q+1} : \mathbf{x}'\mathbf{x} = 1\}$ of $\mathbb{R}^{q+1}$, with $q \geq 1$, is one of the first steps when analysing *directional data*, that is, data supported on Ω_q. Directional data arise in many applied disciplines, such as astronomy or biology, and have been the focus of a considerable number of monographs; see, e.g., [10, 11]. Since the seminal paper by Lord Rayleigh [13], and despite its relative concreteness, the century-old topic of testing uniformity on Ω_q has attracted more than 30 proposals of tests with varying

E. García-Portugués (✉)
Department of Statistics, Carlos III University of Madrid, Leganés, Spain
e-mail: edgarcia@est-econ.uc3m.es

P. Navarro-Esteban · J. A. Cuesta-Albertos
Department of Mathematics, Statistics and Computer Science, University of Cantabria, Santander, Spain
e-mail: paula.navarro@unican.es

J. A. Cuesta-Albertos
e-mail: juan.cuesta@unican.es

S. Balzano et al. (eds.), *Statistical Learning and Modeling in Data Analysis*, Studies in Classification, Data Analysis, and Knowledge Organization,
https://doi.org/10.1007/978-3-030-69944-4_12

degrees of generality (many are circular- or spherical-specific tests, i.e., they assume $q = 1$ or $q = 2$); see [6] for a review on the topic.

Testing uniformity on Ω_2 has several applications in astronomy. An instance is the analysis of the presumed uniform orbit distribution of long-period comets originating in the nearly isotropic Oort cloud [1]. Another application is in the analysis of the distribution of crater impacts, a valuable informer on the impactors that create them. For instance, the case study in [8] for Rhea attributes the uniform-like distributions of small craters to the predominance of planet-orbiting impactors caused by returning debris ejected from large crater impacts. Sun-orbiting impactors, on the other hand, tend to be related to non-uniform crater distributions.

In this work, we propose yet another test of uniformity on Ω_q. The test is based on projections, it is of a Cramér–von Mises nature, and it has the following main appeals: (*i*) applicability to arbitrary dimensions $q \geq 1$; (*ii*) consistency against any alternative to uniformity, i.e., *omnibusness*; (*iii*) conceptual neat extension of the well-known Watson [17] test of *circular* uniformity; (*iv*) known and usable asymptotic distribution; (*v*) computational tractability for the most common dimensions.

The contents of the work are organized as follows. Section 2 sets the problem (Sect. 2.1), reviews a projection-based test of uniformity that motivates this work (Sect. 2.2), and exposes the projected uniformity distribution (Sect. 2.3). Section 3 presents the new test of uniformity, providing the genesis of the test statistic (Sect. 3.1), its U-statistic form (Sect. 3.2), and its asymptotic null distribution (Sect. 3.3). Numerical experiments given in Sect. 4 evidence the tractability of the asymptotic distribution and the fast convergence of the test statistic towards it. Finally, Sect. 5 investigates whether Venusian craters are uniformly distributed.

2 Background

2.1 Testing Uniformity on Ω_q

Testing the uniformity of a continuous random variable $\mathbf{X} \sim \mathrm{P}$ supported on Ω_q is a simple goodness-of-fit problem. It is formalized as the testing of

$$\mathscr{H}_0 : \mathrm{P} = \nu_q \quad \text{vs.} \quad \mathscr{H}_1 : \mathrm{P} \neq \nu_q \tag{1}$$

from a sample $\mathbf{X}_1, \ldots, \mathbf{X}_n$ of independent and identically distributed observations from P, the distribution of $\mathbf{X}$, and where ν_q denotes the uniform distribution on Ω_q. The probability density function (pdf) of ν_q assigns density ω_q^{-1} to any point on Ω_q, where $\omega_q := 2\pi^{\frac{q+1}{2}} / \Gamma\left(\frac{q+1}{2}\right)$ denotes the surface area of Ω_q, $q \geq 1$.

If $\mathbf{X} \sim \nu_q$, then $\mathbf{X}$ is identically distributed to any rotation of $\mathbf{X}$. This property suggests that any proper test for $\mathscr{H}_0$ must be rotation invariant, in the sense that the obtained test decision should remain invariant if we apply the test to any rotation of the sample. Recall also that, since $\mathscr{H}_0$ is actually a simple hypothesis that completely specifies a distribution, Monte Carlo calibration of any test statistic for problem (1) is conceptually straightforward (though perhaps computationally costly).

2.2 Using Projections for Assessing Uniformity

A projection-based test of uniformity on Ω_q is proposed in [1]. This test is based on Corollary 3.2 in [2] from which it is easily deduced that, under some mild regularity conditions, if

(i) $\mathbf{X}$ and $\mathbf{Y}$ are two d-dimensional random vectors whose distributions are different and
(ii) $\boldsymbol{\gamma}$ is a random vector independent of $\mathbf{X}$ and $\mathbf{Y}$ with distribution absolutely continuous with respect to the Lebesgue measure,

then the distributions of the projections of $\mathbf{X}$ and $\mathbf{Y}$ on the one-dimensional subspace generated by $\boldsymbol{\gamma}$ almost surely differ.

Taking into account that the distribution of the projections coincide if $\mathbf{X} \sim \mathbf{Y}$, we have that testing $\mathscr{H}_0$ is almost surely equivalent to testing $\mathscr{H}_0^{\boldsymbol{\gamma}} : \mathbf{X}'\boldsymbol{\gamma} \sim \Pi_q$, where Π_q is the distribution of $\gamma'\mathbf{U}$ and $\mathbf{U} \sim \nu_q$ (see Sect. 2.3).

The test by [1] proceeds as follows: (*i*) sample $\boldsymbol{\gamma} \sim \nu_q$; (*ii*) reject $\mathscr{H}_0^{\boldsymbol{\gamma}}$, and consequently $\mathscr{H}_0$, for large values of the Kolmogorov–Smirnov statistic

$$\mathrm{KS}_{n,\boldsymbol{\gamma}} := \sup_{-1\leq x\leq 1} |F_{n,\boldsymbol{\gamma}}(x) - F_q(x)|, \tag{2}$$

where $F_{n,\boldsymbol{\gamma}}$ is the empirical cdf of $\mathbf{X}_1'\boldsymbol{\gamma}, \ldots, \mathbf{X}_n'\boldsymbol{\gamma}$ and F_q is the cdf of Π_q.

The test that rejects $\mathscr{H}_0$ for large values of $\mathrm{KS}_{n,\boldsymbol{\gamma}}$ is omnibus and fast to evaluate. However, it is also dependent on $\boldsymbol{\gamma}$, whose selection adds an extra layer of randomness. As proposed in [1], this can be mitigated by considering k random directions and combining the p-values associated with each of the k tests into the test statistic

$$\mathrm{CCF}_{n,\boldsymbol{\gamma}_1,\ldots,\boldsymbol{\gamma}_k} := \min\{p\text{-value}_1, \ldots, p\text{-value}_k\}, \tag{3}$$

which rejects $\mathscr{H}_0$ for small values. The asymptotic distribution of (3) is unknown and has to be calibrated by Monte Carlo (conditionally on the choice of $\boldsymbol{\gamma}_1, \ldots, \boldsymbol{\gamma}_k$).

2.3 Projected Uniform Distribution

The distribution Π_q is fundamental to any projection-based test of uniformity. It does not depend on $\boldsymbol{\gamma}$ and its pdf (see, e.g., [11, p. 167]) is

$$\mathrm{B}\left(\tfrac{1}{2}, \tfrac{q}{2}\right)^{-1}(1-t^2)^{q/2-1}, \quad t \in [-1, 1],$$

where $\mathrm{B}(a, b) := \Gamma(a)\Gamma(b)/\Gamma(a+b)$. Therefore, $(\gamma'\mathbf{U})^2 \sim \mathrm{Beta}\left(\frac{1}{2}, \frac{q}{2}\right)$ and

$$F_q(x) = \mathrm{B}\left(\tfrac{1}{2}, \tfrac{q}{2}\right)^{-1} \int_{-1}^{x} (1-t^2)^{q/2-1}\,\mathrm{d}t = \frac{1}{2}\left\{1 + \mathrm{sign}(x)\mathrm{I}_{x^2}\left(\tfrac{1}{2}, \tfrac{q}{2}\right)\right\},$$

where $\mathrm{I}_x(a, b) := \mathrm{B}(a, b)^{-1} \int_0^x t^{a-1}(1-t)^{b-1}\,\mathrm{d}t, a, b > 0$, is the regularized incomplete beta function. Trivially, $F_1(x) = 1 - \frac{\cos^{-1}(x)}{\pi}$ and $F_2(x) = \frac{x+1}{2}$ for $x \in [-1, 1]$.

3 A New Test of Uniformity

3.1 *Genesis of the Test Statistic*

Motivated by (2), we consider the Cramér–von Mises statistic given by

$$\mathrm{CvM}_{n,q,\gamma} := n \int_{-1}^{1} \left(F_{n,\gamma}(x) - F_q(x)\right)^2 \mathrm{d}F_q(x). \tag{4}$$

Of course, this statistic still has the issue of being dependent on $\boldsymbol{\gamma}$. Rather than drawing several random directions and aggregating afterwards the tests' outcomes as (3) does, our statistic itself gathers information from all the directions on Ω_q: it is defined as the *expectation* of (4) with respect to $\boldsymbol{\gamma} \sim \nu_q$:

$$\mathrm{CvM}_{n,q} := \mathbb{E}_{\boldsymbol{\gamma}}\left[\mathrm{CvM}_{n,q,\boldsymbol{\gamma}}\right] = n \int_{\Omega_q} \left[\int_{-1}^{1} \left\{F_{n,\boldsymbol{\gamma}}(x) - F_q(x)\right\}^2 \mathrm{d}F_q(x)\right] \nu_q(\mathrm{d}\boldsymbol{\gamma}). \tag{5}$$

The test based on (5) rejects $\mathscr{H}_0$ for large values of $\mathrm{CvM}_{n,q}$.

The integration on all possible projection directions within the test statistic, as (5) does, was firstly considered in the regression context by [3], though employing an empirical measure instead of ν_q in (5). In our setting, the choice of ν_q as the distribution of $\boldsymbol{\gamma}$ is canonical, given that it is the only (deterministic) distribution that makes (5) invariant to rotations of the sample.

3.2 *U-statistic Form*

Form (5) is not computationally pleasant: it involves a univariate integral and a more challenging integral on Ω_q. Such level of complexity is undesirable for a test statistic, provided that eventually it may be required to be calibrated by Monte Carlo. In addition, form (5) obfuscates the quadratic structure of the statistic and complicates obtaining its asymptotic distribution. The next result solves these two issues.

Theorem 1 (*U*-statistic form of $\mathrm{CvM}_{n,q}$; [5]) *The statistic* (5) *can be expressed as*

$$\mathrm{CvM}_{n,q} = \frac{2}{n}\sum_{i<j}\psi_q(\cos^{-1}(\mathbf{X}_i'\mathbf{X}_j)) + \frac{3-2n}{6}, \tag{6}$$

where, for $\theta \in [0, \pi]$,

$$\psi_q(\theta) = \begin{cases} \frac{1}{2} + \frac{\theta}{2\pi}\left(\frac{\theta}{2\pi} - 1\right), & q = 1, \\ \frac{1}{2} - \frac{1}{4}\sin\left(\frac{\theta}{2}\right), & q = 2, \\ \psi_1(\theta) + \frac{1}{4\pi^2}\left((\pi - \theta)\tan\left(\frac{\theta}{2}\right) - 2\sin^2\left(\frac{\theta}{2}\right)\right), & q = 3, \\ -\frac{3}{4} + \frac{\theta}{2\pi} + 2F_q^2\left(\cos\left(\frac{\theta}{2}\right)\right) \\ \quad -4\int_0^{\cos(\theta/2)} F_q(t)F_{q-1}\left(\frac{t\tan(\theta/2)}{(1-t^2)^{1/2}}\right)\mathrm{d}F_q(t), & q \geq 4. \end{cases}$$

The proof of Theorem 1 is lengthy and therefore omitted. This is also the case for the rest of the presented results. The reader is referred to [5] for the detailed proofs.

The case $q = 1$ of $\mathrm{CvM}_{n,q}$ is especially interesting. It connects with Watson [17]'s well-known U_n^2 statistic for testing the uniformity of a circular sample, defined as

$$U_n^2 := n\int_0^{2\pi}\left\{F_n(\theta) - F_0(\theta) - \int_0^{2\pi}(F_n(\varphi) - F_0(\varphi))\,\mathrm{d}F_0(\varphi)\right\}^2 \mathrm{d}F_0(\theta),$$

where $F_n(\theta) := \frac{1}{n}\sum_{i=1}^n 1_{\{\Theta_i \leq \theta\}}$ is the empirical cdf of the circular sample $\Theta_1, \ldots, \Theta_n$ in $[0, 2\pi)$ and $F_0(\theta) := \theta/(2\pi)$ is the uniform cdf on $[0, 2\pi)$. The U_n^2 statistic can be regarded as the rotation-invariant version of the Cramér–von Mises statistic for circular data, achieving such invariance by minimizing the discrepancy of the sample with respect to $\mathscr{H}_0$ (see, e.g., [6]).

The relation between U_n^2 and $\mathrm{CvM}_{n,1}$ stems from the following alternative form for U_n^2 (see, e.g., [11, p. 111]):

$$U_n^2 = \frac{1}{n}\sum_{i,j=1}^n h\left(\Theta_{ij}\right), \quad h(\theta) := \frac{1}{2}\left(\frac{\theta^2}{4\pi^2} - \frac{\theta}{2\pi} + \frac{1}{6}\right). \tag{7}$$

Here $\Theta_{ij} := \cos^{-1}(\cos(\Theta_i - \Theta_j)) \in [0, \pi]$ is the shortest angle distance between Θ_i and Θ_j. Therefore, if we denote by $\Theta_1, \ldots, \Theta_n$ the angles determining the sample $\mathbf{X}_1, \ldots, \mathbf{X}_n$, it happens that $\cos^{-1}(\mathbf{X}_i'\mathbf{X}_j) = \Theta_{ij}$. From this point, elaborating on the expressions in Theorem 1 when $q = 1$ leads to $\mathrm{CvM}_{n,1} = \frac{1}{2}U_n^2$. Therefore, our claim that the test based on $\mathrm{CvM}_{n,q}$ is an extension of the Watson test to Ω_q, as stated in the following corollary.

Corollary 1 (An extension of the Watson test to Ω_q) *It happens that* $\mathrm{CvM}_{n,1} = \frac{1}{2}U_n^2$. *Consequently, the test that rejects for large values of* $\mathrm{CvM}_{n,1}$ *is equivalent to the Watson test.*

3.3 Asymptotic Distribution

Expression (6) unveils the U-statistic nature of $\mathrm{CvM}_{n,q}$. Since the U-statistic can be seen to be degenerate, the asymptotic distribution of $\mathrm{CvM}_{n,q}$ is an infinite weighted sum of chi-squared random variables. It involves the coefficients $\{b_{k,q}\}$ such that

$$b_{k,q} = \begin{cases} \frac{2}{\pi} \int_0^\pi \psi_1(\theta) T_k(\cos\theta)\, \mathrm{d}\theta, & q = 1, \\ \frac{1}{c_{k,q}} \int_0^\pi \psi_q(\theta) C_k^{(q-1)/2}(\cos\theta) \sin^{q-1}(\theta)\, \mathrm{d}\theta, & q \geq 2, \end{cases}$$

where T_k represents the kth Chebyshev polynomial of the first kind, $C_k^{(q-1)/2}$ stands for the kth Gegenbauer polynomial of order $(q-1)/2$, and

$$c_{k,q} := \frac{2^{3-q} \pi \Gamma(q+k-1)}{(q+2k-1) k! \Gamma((q-1)/2)^2}.$$

Theorem 2 (Asymptotic null distribution; [5]) *Under $\mathscr{H}_0$ and for $q \geq 1$,*

$$\mathrm{CvM}_{n,q} \overset{d}{\rightsquigarrow} \begin{cases} \frac{1}{2} \sum_{k=1}^\infty b_{k,1} \chi^2_{d_{k,1}}, & q = 1, \\ \sum_{k=1}^\infty \frac{q-1}{q-1+2k} b_{k,q} \chi^2_{d_{k,q}}, & q \geq 2, \end{cases} \tag{8}$$

where $\chi^2_{d_{k,q}}$, $k \geq 1$, are independent chi-squared random variables with degrees of freedom

$$d_{k,q} := \binom{q+k-2}{q-1} + \binom{q+k-1}{q-1}.$$

The coefficients $\{b_{k,q}\}$ are non-negative and satisfy $\sum_{k=1}^\infty b_{k,q} d_{k,q} < \infty$.

The coefficients $\{b_{k,q}\}$ admit explicit expressions that drastically improve the tractability of the asymptotic null distribution of $\mathrm{CvM}_{n,q}$ for all $q \geq 1$.

Theorem 3 (Coefficients for ψ_q; [5]) *Let $k \geq 1$. For $q \geq 1$,*

$$b_{k,q} = \begin{cases} \frac{1}{\pi^2 k^2}, & q = 1, \\ \frac{1}{2(2k+3)(2k-1)}, & q = 2, \\ \frac{35}{72\pi^2} 1_{\{k=1\}} + \frac{1}{2\pi^2} \frac{3k^2+6k+4}{k^2(k+1)(k+2)^2} 1_{\{k>1\}}, & q = 3, \\ \frac{(q-1)^2 (2k+q-1) \Gamma((q-1)/2)^3 \Gamma(3q/2)}{8\pi q^2 \Gamma(q/2)^3 \Gamma((3q+1)/2)} & \\ \quad \times {}_4F_3\left(1-k, q+k, \frac{q+1}{2}, \frac{3q}{2}; q+1, \frac{q}{2}+1, \frac{3q+1}{2}; 1\right), & q \geq 4, \end{cases}$$

where ${}_4F_3$ stands for the generalized hypergeometric function.

The final result is a consequence of the fact that $b_{k,q} > 0$, for all $k \geq 1$ and $q \geq 1$, and the fact that $\mathrm{CvM}_{n,q}$ belongs to the class of Sobolev tests [7].

Corollary 2 (Omnibusness) *The test that rejects* $\mathscr{H}_0$ *for large values of* $\mathrm{CvM}_{n,q}$ *is consistent against all alternatives to uniformity with square-integrable pdf.*

4 Numerical Experiments

The asymptotic distributions (8) are usable in practice. The closed forms of $\{b_{k,q}\}$ and the (exact) Imhof [9]'s method allow to compute asymptotic p-values through the evaluation of the truncated-series tail probability function:

$$x \mapsto \mathbb{P}\Big[\sum_{k=1}^{K} w_{k,q}\chi^2_{d_{k,q}} > x\Big] \tag{9}$$

where $x \geq 0$ and K is a "sufficiently large" integer. Asymptotic critical values c_α for a significance level α are computable using a numerical inversion on (9).

The first numerical experiment investigates how large K must be for ensuring a uniform error bound in (9), relatively to $K = 10^5$. Figure 1 evidences that (9) converges slower, as a function of K, for increasing q's. It also gives simple takeaways: (*i*) $K = 10^3$ ensures asymptotic p-values with uniform error bound $\epsilon = 5 \times 10^{-3}$ for $q \leq 10$; (*ii*) $K = 10^4$ decreases the uniform error bound to $\epsilon = 5 \times 10^{-4}$; (*iii*) the accuracy for lower p-values, approximately in the [0, 0.15]-range (left side of the horizontal axis), improves over the uniform bound.

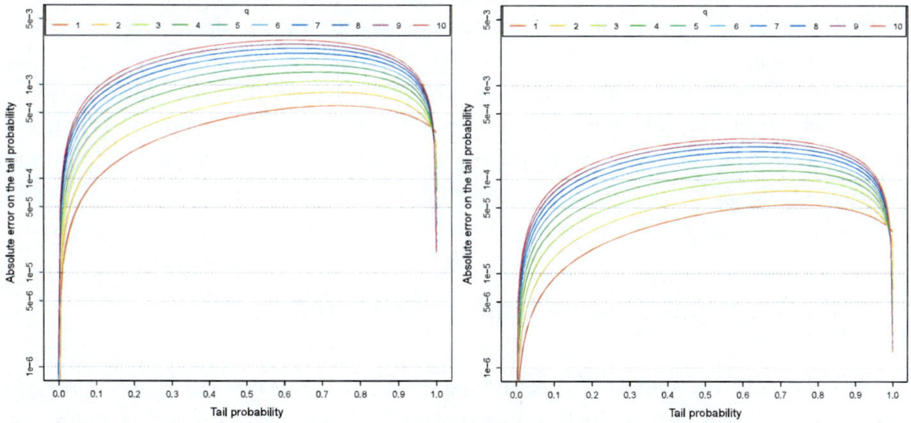

Fig. 1 Accuracy of the truncation of (9), computed with Imhof's method. The vertical axis shows the absolute errors, with respect to $K = 10^5$, of considering $K = 10^3$ (left) and $K = 10^4$ (right). The horizontal axis shows the probability of (9) with $K = 10^3$ (a common [0, 1]-scale for all curves)

Table 1 Critical values of the $\mathrm{CvM}_{n,q}$ statistic, approximated by $M = 10^6$ Monte Carlo replicates. The asymptotic (∞) critical values result from computing and inverting (9) with $K = 10^4$

α	n	q									
		1	2	3	4	5	6	7	8	9	10
0.10											
	25	0.3015	0.2752	0.2588	0.2481	0.2401	0.2343	0.2295	0.2256	0.2223	0.2193
	50	0.3026	0.2760	0.2600	0.2490	0.2411	0.2351	0.2302	0.2264	0.2229	0.2201
	100	0.3029	0.2765	0.2605	0.2496	0.2416	0.2355	0.2307	0.2268	0.2234	0.2206
	200	0.3032	0.2769	0.2608	0.2498	0.2419	0.2357	0.2309	0.2270	0.2236	0.2207
	400	0.3036	0.2769	0.2608	0.2502	0.2423	0.2360	0.2311	0.2272	0.2237	0.2209
	∞	0.3035	0.2769	0.2607	0.2498	0.2419	0.2358	0.2309	0.2269	0.2236	0.2207
0.05											
	25	0.3696	0.3254	0.2994	0.2824	0.2703	0.2613	0.2541	0.2483	0.2434	0.2394
	50	0.3716	0.3273	0.3012	0.2841	0.2719	0.2627	0.2554	0.2495	0.2446	0.2403
	100	0.3730	0.3284	0.3027	0.2852	0.2730	0.2635	0.2563	0.2503	0.2453	0.2411
	200	0.3728	0.3290	0.3029	0.2857	0.2734	0.2638	0.2566	0.2506	0.2456	0.2414
	400	0.3744	0.3288	0.3029	0.2859	0.2735	0.2639	0.2566	0.2508	0.2457	0.2417
	∞	0.3737	0.3291	0.3029	0.2856	0.2733	0.2639	0.2566	0.2506	0.2456	0.2414
0.01											
	25	0.5220	0.4360	0.3868	0.3561	0.3349	0.3186	0.3062	0.2958	0.2876	0.2805
	50	0.5306	0.4412	0.3920	0.3601	0.3384	0.3219	0.3090	0.2983	0.2903	0.2830
	100	0.5339	0.4451	0.3948	0.3626	0.3400	0.3235	0.3105	0.3002	0.2915	0.2842
	200	0.5359	0.4467	0.3962	0.3642	0.3405	0.3238	0.3112	0.3006	0.2916	0.2843
	400	0.5368	0.4463	0.3968	0.3635	0.3409	0.3242	0.3114	0.3006	0.2921	0.2849
	∞	0.5368	0.4469	0.3963	0.3639	0.3413	0.3244	0.3113	0.3008	0.2921	0.2848

The second numerical experiment evaluates the convergence speed of (8) with Table 1, which gives the critical values of the statistic for dimensions $q = 1, \ldots, 10$ and significance levels $\alpha = 0.10, 0.05, 0.01$. As it is unveiled, the convergence towards the asymptotic distribution is quite fast, for all the dimensions explored, effectively requiring to save a single critical value for each dimension q to yield a test decision. The critical values steadily decrease with the increment of the dimension.

5 Are Venusian Craters Uniformly Distributed?

Venus is the closest planet to Earth and the most Earth-like planet of the Solar System in terms of size and composition. As such, it is one of the most explored extraterrestrial bodies by humankind, a landmark on its exploration being the Magellan mission (1989–1994). Through a series of mapping cycles, the Magellan spacecraft produced the first global, high-resolution mapping of 98% of the Venusian surface. The analysis

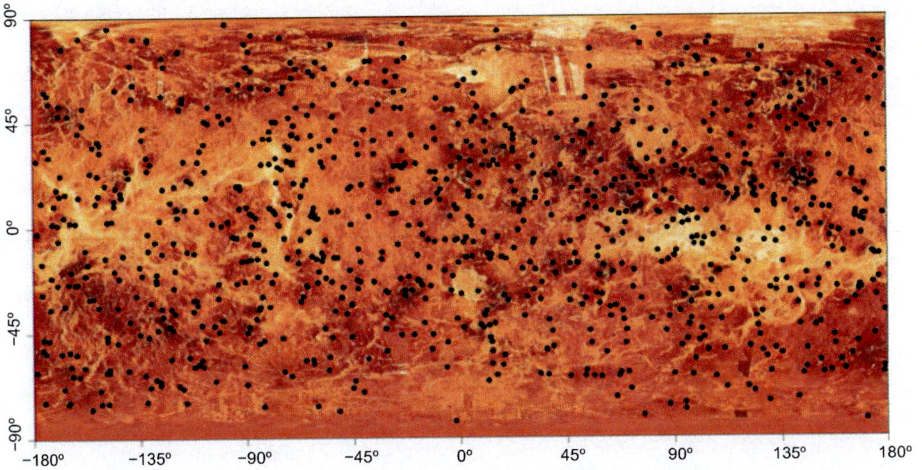

Fig. 2 Venusian craters (black points) overlaid over a colourized image of the Venus surface [16]

of the vast imagery produced in the mission (see [4]) revealed the high uniformity of the Venusian crater distribution [12, 14]. Indeed, [12, Sect. 3.1] tested the uniformity of such distribution using the 763 locations of craters back then available, finding no evidence to reject $\mathscr{H}_0$ for several tests.

We tested uniformity with an updated database of Venusian craters [15] that contains the locations of the centres of 967 craters. Figure 2 shows these locations over Venus' surface, as mapped by the Magellan mission. We performed the Rayleigh [13] and Giné's F_n [7] tests, as considered by [14], the Cuesta-Albertos et al. [1] test (using $k = 50$), and the novel $\mathrm{CvM}_{n,2}$-based test. The obtained p-values, estimated with 10^4 Monte Carlo replicates, were 0.170, 0.112, 0.117, and 0.129, respectively. Consequently, we found no statistical evidence to reject $\mathscr{H}_0$ at usual significance levels, thus confirming the analysis by [12] with updated crater records.

The apparent uniform distribution of Venusian craters is truly remarkable. Indeed, among the very few planets and moons of the Solar System with uniformly distributed craters, Venus has the largest number of observed craters, according to the database of named craters of the International Astronomical Union [5]. The filtering of small meteoroids by the dense Venusian atmosphere may be one of the causes explaining the uniform distribution of craters.

Acknowledgements The first author acknowledges financial support from grants PGC2018-097284-B-I00, IJCI-2017-32005 and MTM2016-76969-P, funded by the Spanish Ministry of Economy, Industry and Competitiveness, and the European Regional Development Fund. The second and third authors acknowledge financial support from grant MTM2017-86061-C2-2-P from the Spanish Ministry of Economy, Industry and Competitiveness. The authors gratefully acknowledge the computing resources of the Supercomputing Center of Galicia (CESGA). Comments by two referees are acknowledged.

References

1. Cuesta-Albertos, J.A., Cuevas, A., Fraiman, R.: On projection-based tests for directional and compositional data. Stat. Comput. **19**(4), 367–380 (2009). https://doi.org/10.1007/s11222-008-9098-3
2. Cuesta-Albertos, J.A., Fraiman, R., Ransford, T.: A sharp form of the Cramér-Wold theorem. J. Theor. Probab. **20**(4), 201–209 (2007). https://doi.org/10.1007/s10959-007-0060-7
3. Escanciano, J.C.: A consistent diagnostic test for regression models using projections. Econ. Theory. **22**(6), 1030–1051 (2006). https://doi.org/10.1017/S0266466606060506
4. Ford, J.P., Plaut, J.J., Weitz, C.M., Farr, T.G., Senske, D.A., Stofan, E.R., Michaels, G., Parker, T.J.: Guide to Magellan image interpretation. Tech. Rep. JPL Publication **93–24**, (1993)
5. García-Portugués, E., Navarro-Esteban, P., Cuesta-Albertos, J.A.: On a projection-based class of uniformity tests on the hypersphere (2020). arXiv:2008.09897
6. García-Portugués, E., Verdebout, T.: An overview of uniformity tests on the hypersphere (2018). arXiv:1804.00286
7. Giné, E.: Invariant tests for uniformity on compact Riemannian manifolds based on Sobolev norms. Ann. Stat. **3**(6), 1243–1266 (1975). https://doi.org/10.1214/aos/1176343283
8. Hirata, N.: Differential impact cratering of Saturn's satellites by heliocentric impactors. J. Geophys. Res. Planets **121**(2), 111–117 (2016). https://doi.org/10.1002/2015JE004940
9. Imhof, J.P.: Computing the distribution of quadratic forms in normal variables. Biometrika **48**(3/4), 419–426 (1961). https://doi.org/10.1093/biomet/48.3-4.419
10. Ley, C., Verdebout, T.: Modern Directional Statistics. CRC press, Boca Raton (2017). https://doi.org/10.1201/9781315119472
11. Mardia, K.V., Jupp, P.E.: Directional Statistics. Wiley, Chichester (1999). https://doi.org/10.1002/0471667196.ess7086
12. Phillips, R.J., Raubertas, R.F., Arvidson, R.E., Sarkar, I.C., Herrick, R.R., Izenberg, N., Grimm, R.E.: Impact craters and Venus resurfacing history. J. Geophys. Res. **97**(E10), 15923–15948 (1992). https://doi.org/10.1029/92JE01696
13. Rayleigh, Lord: On the problem of random vibrations, and of random flights in one, two, or three dimensions. Lond. Edinb. Dublin Philos. Mag. J. Sci. **37**(220), 321–347 (1919). https://doi.org/10.1080/14786440408635894
14. Schaber, G.G., Strom, R.G., Moore, H.J., Soderblom, L.A., Kirk, R.L., Chadwick, D.J., Dawson, D.D., Gaddis, L.R., Boyce, J.M., Russell, J.: Geology and distribution of impact craters on Venus: What are they telling us? J. Geophys. Res. **97**(E8), 13257–13301 (1992). https://doi.org/10.1029/92JE01246
15. USGS Astrogeology Science Center: Venus Crater Database (2011). https://astrogeology.usgs.gov/search/map/Venus. Accessed January 23, 2020
16. USGS Astrogeology Science Center: Venus Magellan Global C3-MDIR Synthetic Color Mosaic 4641m v1 (2014). https://astrogeology.usgs.gov/search/map/Venus/Magellan/Colorized/Venus. Accessed January 23, 2020
17. Watson, G.S.: Goodness-of-fit tests on a circle. Biometrika **48**(1/2), 109–114 (1961). https://doi.org/10.2307/2333135

On Mean And/or Variance Mixtures of Normal Distributions

Sharon X. Lee and Geoffrey J. McLachlan

Abstract Parametric distributions are an important part of statistics. There is now a voluminous literature on different fascinating formulations of flexible distributions. We present a selective and brief overview of a small subset of these distributions, focusing on those that are obtained by scaling the mean and/or covariance matrix of the (multivariate) normal distribution with some scaling variable(s). Namely, we consider the families of the mean mixture, variance mixture, and mean–variance mixture of normal distributions. Their basic properties, some notable special/limiting cases, and parameter estimation methods are also described.

Keywords Scale mixture distribution · Mean mixture distribution · Non-normal distribution · Normal distribution · Skew-normal distribution

1 Introduction

The normal distribution plays a central role in statistical modeling and data analysis, but real data rarely follow this classical distribution. The quest for more flexible distributions has led to an ever-growing development in the literature of parametric distributions. In the past two decades or so, intense interest has been in the area of skew or asymmetric distributions; see, for example, the book edited by [18], the monograph by [10], and the papers by [2, 5, 9] for recent accounts of the literature on skew distributions. Many of these formulations belong to the class of skew-symmetric distributions, which is a generalization of the classical skew-normal (SN) distribution by [11]. This SN distribution can be characterized as a mean mixture of normal (MMN) distribution, where the mean of a normal random variable is scaled

S. X. Lee (✉)
University of Adelaide, Adelaide, Australia
e-mail: sharon.lee@adelaide.edu.au

G. J. McLachlan
University of Queensland, Brisbane, Australia
e-mail: g.mclachlan@uq.edu.au

S. Balzano et al. (eds.), *Statistical Learning and Modeling in Data Analysis*, Studies in Classification, Data Analysis, and Knowledge Organization,
https://doi.org/10.1007/978-3-030-69944-4_13

by a truncated normal random variable [28]. Another related and extensively studied family of distributions that can render asymmetric distributional shapes is the mean–variance mixture of normal (MVMN) distribution. Introduced by [12], the MVMN distribution is obtained by scaling both mean and variance of a normal random variable with the same (positive scalar) scaling random variable. These distributions belong to the recently proposed class of generalized mixtures of normal distributions presented by [6].

This paper presents a brief overview of flexible distributions that arise from scaling either/both the mean and variance of a normal random variable. For simplicity, we focus on the case of a univariate scaling variable. Apart from the aforementioned MMN and MVMN families, a third family called variance mixture of normal (VMN) distribution can be defined by scaling only the variance of a normal random variable. Although VMN does not produce asymmetric distributions (at least not in the case of a scalar scaling variable), we include this family in this paper for completeness.

The flexibility to model non-normal data features has rendered these three families of distributions useful in a wide range of applications. In particular, some of the special cases (such as the generalized hyperbolic distribution and the skew-normal distribution) are enjoying increasing popularity in applications ranging from bioinformatics, climatology, and fisheries to finance and social sciences [3, 8, 14, 22, 30, 32].

Following conventional notation, a p-dimensional random vector $\boldsymbol{Y}$ is said to follow a (multivariate) normal distribution, denoted by $\boldsymbol{Y} \sim N_p(\boldsymbol{\mu}, \boldsymbol{\Sigma})$, if its density is given by

$$\phi_p(\boldsymbol{Y}; \boldsymbol{\mu}, \boldsymbol{\Sigma}) = (2\pi)^{-\frac{p}{2}} |\boldsymbol{\Sigma}|^{-\frac{1}{2}} e^{-\frac{1}{2}(\boldsymbol{Y}-\boldsymbol{\mu})^\top \boldsymbol{\Sigma}^{-1}(\boldsymbol{Y}-\boldsymbol{\mu})}, \tag{1}$$

where $\boldsymbol{\mu}$ is a $p \times 1$ vector of location parameters and $\boldsymbol{\Sigma}$ is $p \times p$ positive definite symmetric matrix of scale parameters. The mean and variance of $\boldsymbol{Y}$ are $E(\boldsymbol{Y}) = \boldsymbol{\mu}$ and $\text{cov}(\boldsymbol{Y}) = \boldsymbol{\Sigma}$, respectively. The vector $\boldsymbol{Y}$ can be expressed as a location-scale variant of a standard normal random variable, that is,

$$\boldsymbol{Y} = \boldsymbol{\mu} + \boldsymbol{\Sigma}^{\frac{1}{2}} \boldsymbol{Z}, \tag{2}$$

where $\boldsymbol{Z} \sim N_p(\boldsymbol{0}, \boldsymbol{I}_p)$, $\boldsymbol{0}$ is a vector of zeros, and $\boldsymbol{I}_p$ is the $p \times p$ identity matrix. By 'scaling' or 'mixing' $\boldsymbol{Y}$, we mean that $\boldsymbol{\mu}$ is mixed with W and/or $\boldsymbol{\Sigma}$ is weighted by $\sqrt{W}$, where W is a positive random variable independent of $\boldsymbol{Z}$. We consider each of these cases in Sects. 2–4. By adopting a range of different distributions for W, a wide variety of non-normal distributions can be constructed.

Each of the MMN, VMN, and MVMN models have their own (theoretical and practical) advantages and limitations. For example, some facilitates parameter estimation procedures that are easier to implement and less computationally demanding to compute, while some others may offer nice features such as more flexible distributional shapes.

2 Mean Mixture of Normal Distributions

The mean mixture (or location-mixture) of normal (MMN) distribution [28] refers to the family of distributions generated by mixing the location parameter $\boldsymbol{\mu}$ with a (scalar) variable W. More formally, the MMN distribution arises from the stochastic expression

$$\boldsymbol{Y} = \boldsymbol{\mu} + W\boldsymbol{\delta} + \boldsymbol{\Sigma}^{\frac{1}{2}}\boldsymbol{Z}, \tag{3}$$

where $\boldsymbol{\delta}$ is $p \times 1$ vector of shape parameters. The density (3) is asymmetric if W has an asymmetric distribution. In this case, $\boldsymbol{\delta}$ may be interpreted as a vector of skewness parameters. A prominent example is the (positively) truncated normal or half-normal distribution, that is, $W \sim TN(0, 1; \mathbb{R}^+)$. This leads to the classical characterization of the skew-normal (SN) distribution proposed by [11].

From (3), the density of the MMN distribution can be expressed as

$$f(\boldsymbol{Y}; \boldsymbol{\mu}, \boldsymbol{\Sigma}, \boldsymbol{\delta}; h(w; \boldsymbol{\theta})) = \int_{-\infty}^{\infty} \phi_p(\boldsymbol{Y}; \boldsymbol{\mu} + w\boldsymbol{\delta}, \boldsymbol{\Sigma})\, h(w; \boldsymbol{\theta})dw, \tag{4}$$

where, again, $h(w; \boldsymbol{\theta})$ denotes the density of W with parameter $\boldsymbol{\theta}$. The notation $\boldsymbol{Y} \sim MMN_p(\boldsymbol{\mu}, \boldsymbol{\Sigma}, \boldsymbol{\delta}; h(w; \boldsymbol{\theta}))$ will be used when $\boldsymbol{Y}$ has density in the form of (4). Similar to the VMN distribution, the MMN distribution admits a two-level hierarchical representation given by

$$\boldsymbol{Y}|W = w \sim N_p(\boldsymbol{\mu} + w\boldsymbol{\delta}, \boldsymbol{\Sigma}) \perp W \sim h(w; \boldsymbol{\theta}). \tag{5}$$

2.1 *Properties*

It is straightforward to obtain the moments for a MMN random variable. From the stochastic representation (3), it can be seen that first moment of $\boldsymbol{Y}$ is given by $E(\boldsymbol{Y}) = \boldsymbol{\mu} + E(W)\boldsymbol{\delta}$ if $E(|W|) < \infty$. Similarly, the second moment of $\boldsymbol{Y}$ is given by $\text{cov}(\boldsymbol{Y}) = \boldsymbol{\Sigma} - \text{var}(W)\boldsymbol{\delta}\boldsymbol{\delta}^\top$, provided $E(W^2)$ is finite. Further, the mgf of $\boldsymbol{Y}$ is given by

$$M_{\boldsymbol{Y}}(\boldsymbol{t}) = e^{\boldsymbol{t}^\top\boldsymbol{\mu} + \frac{1}{2}\boldsymbol{t}^\top\boldsymbol{\Sigma}\boldsymbol{t}} M_W\left(\boldsymbol{t}^\top\boldsymbol{\delta}\right). \tag{6}$$

The MMN distribution also enjoys nice properties such as closure under linear transformation, marginalization, and conditioning. Let $\boldsymbol{Y} \sim MMN_p(\boldsymbol{\mu}, \boldsymbol{\Sigma}, \boldsymbol{\delta}; h(w; \boldsymbol{\theta})$, $\boldsymbol{A}$ be a $q \times p$ matrix of full row rank, and $\boldsymbol{a}$ be a q-dimensional vector. Then the affine transformation $\boldsymbol{AY} + \boldsymbol{a}$ still has a NMM distribution, given by

$$\boldsymbol{AY} + \boldsymbol{a} \sim MMN_q(\boldsymbol{A\mu} + \boldsymbol{b}, \boldsymbol{A\Sigma A}^\top, \boldsymbol{A\delta}; h(w; \boldsymbol{\theta})). \tag{7}$$

In addition, the linear combination of a MMN and a normal random variables is also a MMN random variable. If $\boldsymbol{X} \sim N_q(\boldsymbol{\mu}^*, \boldsymbol{\Sigma}^*)$ is independent of $\boldsymbol{Y}$, then the linear combination $\boldsymbol{AY} + \boldsymbol{X}$ is distributed as

$$\boldsymbol{AY} + \boldsymbol{X} \sim MMN_q(\boldsymbol{A\mu} + \boldsymbol{\mu}^*, \boldsymbol{A\Sigma A}^\top + \boldsymbol{\Sigma}^*, \boldsymbol{A\delta}; h(w; \boldsymbol{\theta})). \tag{8}$$

Concerning the marginal and conditional distributions of MMN random variables, let $\boldsymbol{Y}$, $\boldsymbol{\mu}$, and $\boldsymbol{\Sigma}$ be partitioned as in Sect. 4.1. Similarly, partition $\boldsymbol{\delta}$ into $\boldsymbol{\delta}^\top = (\boldsymbol{\delta}_1^\top, \boldsymbol{\delta}_2^\top)$. Then the marginal density of $\boldsymbol{Y}_1$ is $MMN_{p_1}(\boldsymbol{\mu}_1, \boldsymbol{\Sigma}_{11}, \boldsymbol{\delta}_1; h(w; \boldsymbol{\theta}))$ and the conditional density of $\boldsymbol{Y}_1|\boldsymbol{Y}_2 = \boldsymbol{Y}_2$ is $MMN_{p_1}(\boldsymbol{\mu}_{1.2}, \boldsymbol{\Sigma}_{11.2}, \boldsymbol{\delta}_{1.2}; h(w; \boldsymbol{\theta}))$, where $\boldsymbol{\delta}_{1.2} = \boldsymbol{\delta}_1 - \boldsymbol{\Sigma}_{12}\boldsymbol{\Sigma}_{22}^{-1}\boldsymbol{\delta}_2$, and $\boldsymbol{\mu}_{1.2}$ and $\boldsymbol{\Sigma}_{11.2}$ are defined in Sect. 4.1.

2.2 *Special Cases*

As mentioned previously, taking $W \sim TN(0, 1; \mathbb{R}^+)$ leads to the classical SN density given by

$$f(\boldsymbol{Y}; \boldsymbol{\mu}, \boldsymbol{\Sigma}, \boldsymbol{\delta}) = 2\,\phi_p(\boldsymbol{Y}; \boldsymbol{\mu}, \boldsymbol{\Omega})\Phi_1(\boldsymbol{\delta}^\top\boldsymbol{\Omega}^{-1}(\boldsymbol{Y} - \boldsymbol{\mu}); 0, 1 - \boldsymbol{\delta}^\top\boldsymbol{\Omega}^{-1}\boldsymbol{\delta}), \tag{9}$$

where $\boldsymbol{\Omega} = \boldsymbol{\Sigma} + \boldsymbol{\delta\delta}^\top$ and $\Phi_1(\cdot; \mu, \sigma^2)$ denotes the corresponding distribution function of $\phi_1(\cdot; \mu, \sigma^2)$. When $\boldsymbol{\delta} = \boldsymbol{0}$, the SN distribution reduces to the (multivariate) normal distribution.

Another special case of the MMN distribution was presented in [28]. Taking W to have a standard exponential distribution, that is, $W \sim \exp(1)$, leads to the MMN exponential (MMNE) distribution. It can be shown that the density is given by

$$f(\boldsymbol{Y}; \boldsymbol{\mu}, \boldsymbol{\Sigma}, \boldsymbol{\delta}) = \frac{\sqrt{2\pi}}{\alpha} e^{\frac{\beta^2}{2}}\, \Phi_p(\boldsymbol{Y}; \boldsymbol{\mu}, \boldsymbol{\Sigma})\, \Phi_1(\beta), \tag{10}$$

where $\alpha^2 = \boldsymbol{\delta}^\top\boldsymbol{\Sigma}^{-1}\boldsymbol{\delta}$ and $\beta = \alpha^{-1}\left[\boldsymbol{\delta}^\top\boldsymbol{\Sigma}^{-1}(\boldsymbol{Y} - \boldsymbol{\mu}) - 1\right]$. For further details and properties of the MMNE distribution, the reader is referred to Sect. 8.1 in [28]. The flexibility of the MMN and MMNE distributions has been demonstrated applications such as the modeling of environmental data [28] and segmentation of satellite images [27].

3 Mean–Variance Mixture of Normal Distributions

The mean–variance mixture of normal (MVMN) distribution, sometimes called the location-scale mixture of normal distribution, is a generalization of the VMN distribution described in Sect. 2. Compared to (3), the scaling parameter $\boldsymbol{\Sigma}$ is now also weighted with W like in the case of the MVN distribution. The MVMN distribution has the following stochastic representation:

$$\boldsymbol{Y} = \boldsymbol{\mu} + W\boldsymbol{\delta} + \sqrt{W}\boldsymbol{\Sigma}^{\frac{1}{2}}\boldsymbol{Z}. \tag{11}$$

In this case, both the location and scale of the distribution vary with W. Moreover, W is a positive random variable and hence the MVMN distribution is asymmetric when $\boldsymbol{\delta} \neq \boldsymbol{0}$. It is important to note that while the MVMN distribution reduces to the VMN distribution when $\boldsymbol{\delta} = \boldsymbol{0}$, the MMN distribution described in Sect. 2 is not a special case of the MVMN distribution.

Following the definition (11), the density of p-dimensional MVMN distribution can be expressed as

$$f(\boldsymbol{Y}; \boldsymbol{\mu}, \boldsymbol{\Sigma}, \boldsymbol{\delta}; h(w; \boldsymbol{\theta})) = \int_0^\infty \phi_p(\boldsymbol{Y}; \boldsymbol{\mu} + w\delta, w\boldsymbol{\Sigma})\, h(w; \boldsymbol{\theta})dw. \tag{12}$$

The notation $\boldsymbol{Y} \sim MVMN_p(\boldsymbol{\mu}, \boldsymbol{\Sigma}, \boldsymbol{\delta}; h(w; \boldsymbol{\theta}))$ will be used when $\boldsymbol{Y}$ has density in the form of (12). Analogous to the VMN and MMN distributions, the MVMN distribution can be conveniently expressed in a hierarchical form given by

$$\boldsymbol{Y}|W = w \sim N_p(\boldsymbol{\mu} + w\boldsymbol{\delta}, w\boldsymbol{\Sigma}) \perp W \sim h(w; \boldsymbol{\theta}). \tag{13}$$

3.1 Properties

Some basic properties of the MVMN distribution have been studied in [12], among other works. The moments of $\boldsymbol{Y} \sim MVMN_p(\boldsymbol{\mu}, \boldsymbol{\Sigma}, \boldsymbol{\delta}; h(w; \boldsymbol{\theta}))$ can be derived directly from (11). Specifically, the first two moments of $\boldsymbol{Y}$ are given by $E(\boldsymbol{Y}) = \boldsymbol{\mu} + E(W)\boldsymbol{\delta}$ and $\text{cov}(\boldsymbol{Y}) = \text{var}(W)\boldsymbol{\delta}\boldsymbol{\delta}^\top + E(W)\boldsymbol{\Sigma}$, respectively. Further, the mgf of $\boldsymbol{Y}$ is given by

$$M_Y(\boldsymbol{t}) = e^{\boldsymbol{t}^\top \boldsymbol{\mu}} M_W\left(\boldsymbol{t}^\top \delta + \frac{1}{2}\boldsymbol{t}^\top \boldsymbol{\Sigma}\boldsymbol{t}\right). \tag{14}$$

As can be expected, the MVMN distribution shares certain nice properties with the VMN distribution such as closure under linear transformation and marginalization. Let $\boldsymbol{A}$ be a $q \times p$ matrix of full row rank and $\boldsymbol{a}$ be a q-dimensional vector. Then the affine transformation $\boldsymbol{AY} + \boldsymbol{a}$ remains a MVMN distribution. Specifically,

$$\boldsymbol{AY} + \boldsymbol{a} \sim MVMN_q(\boldsymbol{A\mu} + \boldsymbol{b}, \boldsymbol{A\Sigma A}^\top, \boldsymbol{A\delta}; h(w; \boldsymbol{\theta})). \tag{15}$$

Similar to the MMN distribution, a linear combination of a MVMN and a normal random variable remains a MVMN random variable. If $\boldsymbol{X} \sim N_q(\boldsymbol{\mu}^*, \boldsymbol{\Sigma}^*)$ is independent of $\boldsymbol{Y}$, then the linear combination $\boldsymbol{AY} + \boldsymbol{X}$ is distributed as

$$AY + X \sim MVMN_q(A\mu + \mu^*, A\Sigma A^\top + \Sigma^*, A\delta; h(w; \theta)). \tag{16}$$

Marginal distributions and conditional distributions of MVMN random variables can also be derived. Let Y, μ, Σ, and δ be partitioned as in Sect. 2.1. Then the marginal density of Y_1 is $MVMN_{p_1}(\mu_1, \Sigma_{11}, \delta_1; h(w; \theta))$ and the conditional density of $Y_1|Y_2 = Y_2$ is $MVMN_{p_1}(\mu_{1.2}, \Sigma_{11.2}, \delta_{1.2}; h(w; \theta))$, where $\delta_{1.2}$, $\mu_{1.2}$, and $\Sigma_{11.2}$ are defined in Sect. 2.1.

3.2 Special Cases

Perhaps the most well-known special case of the MVMN distribution is the generalized hyperbolic (GH) distribution, which is widely applied in finance and other fields. This distribution is obtained by letting $W \sim GIG(\psi, \chi, \lambda)$, yielding the following density [25]:

$$f(Y; \mu, \Sigma, \delta, \psi, \chi, \lambda) = \frac{\left(\frac{\psi}{\chi}\right)^{\frac{\lambda}{2}} K_{\lambda - \frac{p}{2}}\left(\sqrt{(\psi + d_\delta)(\chi + d_Y)}\right)}{(2\pi)^{\frac{p}{2}} |\Sigma|^{\frac{1}{2}} K_\lambda(\chi\psi) e^{\delta^\top \Sigma^{-1}(Y-\mu)}} \left(\frac{\chi + d_Y}{\psi + d_\delta}\right)^{\frac{\lambda}{2} - \frac{p}{4}}, \tag{17}$$

where $d_\delta = \delta^\top \Sigma^{-1} \delta$, $d_Y = (Y - \mu)^\top \Sigma^{-1} (Y - \mu)$, and $K_\lambda(\cdot)$ denotes the modified Bessel function of the third kind with index λ. Here GIG refers to the generalized inverse gamma distribution. The GH distribution, as the name suggests, contains the symmetric GH distribution mentioned in Sect. 4.2 and an asymmetric version of some of its members. However, the SN distribution cannot be obtained as a special/limiting case. Other noteworthy special cases of the GH distribution include the normal inverse Gaussian, variance gamma, and asymmetric Laplace distributions. The GH distribution and its properties have been well studied in the literature; see, for example, [16, 21].

Two other less well-known MVMN distributions were recently considered by [26, 29]. The former presented a MVMN of Birnbaum–Saunders (MVNBS) distribution, where W has a Birnbaum–Saunders distribution with shape parameter α and scale parameter 1. In the second reference, the authors assumed W follows a Lindley distribution, which is a mixture of $\exp(\alpha)$ and $\text{gamma}(2, \alpha)$ distributions. This leads to the so-called MVN Lindley (MVNL) distribution.

4 Variance Mixture of Normal Distributions

Rather than mixing μ with W, the variance mixture (or scale mixture) of normal (VMN) distribution is obtained by weighting Σ with W. Note that in this case W needs to be positive. More formally, VMN refers to distributions with the following stochastic representation:

$$Y = \mu + \sqrt{W}\Sigma^{\frac{1}{2}}Z, \tag{18}$$

where $Z \sim N(\mathbf{0}, I_p)$ and W are independent. Let the density of W be denoted by $h(w; \theta)$, where θ is the vector of parameters associated with W. It follows that the density is in the form of an integral given by

$$f(Y; \mu, \Sigma, \theta) = \int_0^\infty \phi_p(Y; \mu, w\Sigma)\, h(w; \theta) dw. \tag{19}$$

A similar expression to (19) above can be given in the case where W has a discrete distribution; see, for example, Eq. (3) of [24]. As can be observed from (18), the family of VMN distributions has constant mean but variable scale depending on W. This allows the VM distributions to have lighter or heavier tails than the normal distribution and thus are suitable for modeling data with tails thickness that deviates from the normal. However, this distribution in the unimodal family remains symmetric in shape.

4.1 Properties

The moments of VMN distributions can be readily obtained from (18). For example, the first and second moments of Y are given by, respectively, $E(Y) = \mu$ and $\mathrm{cov}(Y) = E(W)\Sigma$. Further, the moment generating function (mgf) of Y can be expressed as

$$M_Y(t) = e^{t^\top \mu} M_W\left(\frac{1}{2}t^\top \Sigma t\right), \tag{20}$$

where $M_W(\cdot)$ denotes the mgf of W.

Some nice properties of the normal distribution remain valid for VMN distributions, including closure under affine transformation, marginalization, and conditioning. Let $Y \sim VMN_p(\mu, \Sigma; h(w; \theta))$ denotes Y having the density (19). Let also A be a $q \times p$ matrix of full row rank and a be a q-dimensional vector. Then the affine transformation $AY + a$ still has a VMN distribution given by

$$AY + a \sim VMN_q(A\mu + b, A\Sigma A^\top; h(w; \theta)). \tag{21}$$

Furthermore, if $X \sim VMN_q(\mu^*, \Sigma^*; h(w; \theta))$ is independent of Y, then the linear combination $AY + X$ has distribution given by

$$AY + X \sim VMN_q(A\mu + \mu^*, A\Sigma A^\top + \Sigma^*; h(w; \theta)). \tag{22}$$

Suppose Y can be partitioned as $Y^\top = (Y_1^\top, Y_2^\top)$ with respective dimensions p_1 and p_2 where $p_1 + p_2 = p$. Accordingly, let $\mu^\top = (\mu_1^\top, \mu_2^\top)$ and Σ be partitioned into four block matrices Σ_{11}, Σ_{12}, Σ_{21}, and Σ_{22}. Then the marginal density of

$\boldsymbol{Y}_1$ is $VMN_{p_1}(\boldsymbol{\mu}_1, \boldsymbol{\Sigma}_{11}; h(w; \boldsymbol{\theta}))$ and the conditional density of $\boldsymbol{Y}_1|\boldsymbol{Y}_2 = \boldsymbol{Y}_2$ is $VMN_{p_1}(\boldsymbol{\mu}_{1.2}, \boldsymbol{\Sigma}_{11.2}; h(w; \boldsymbol{\theta}))$, where $\boldsymbol{\mu}_{1.2} = \boldsymbol{\mu}_1 + \boldsymbol{\Sigma}_{11}\boldsymbol{\Sigma}_{22}^{-1}(\boldsymbol{Y}_2 - \boldsymbol{\mu}_2)$ and $\boldsymbol{\Sigma}_{11.2} = \boldsymbol{\Sigma}_{11} - \boldsymbol{\Sigma}_{12}\boldsymbol{\Sigma}_{22}^{-1}\boldsymbol{\Sigma}_{21}$.

4.2 Special Cases

The family of VMN distribution encompasses many well-known distributions, including the t, Cauchy, symmetric generalized hyperbolic, and logistic distributions. The slash, Pearson type VII, contaminated normal, and exponential power distributions can also be represented as a VMN distribution; see also [4, 24] for some other special cases of VMN distributions.

The (p-dimensional) t-distribution can be obtained by letting $W \sim IG\left(\frac{\nu}{2}, \frac{\nu}{2}\right)$ in (18), where $IG(\cdot)$ denotes the inverse gamma distribution and ν is a scalar parameter commonly known as the degrees of freedom. This tuning parameter regulates the thickness of the tails of the t-distribution, allowing it to model heavier tails than the normal distribution. The Cauchy and normal distributions are special/limiting cases of the t-distribution (by letting $\nu = 1$ and $\nu \to \infty$, respectively).

The (symmetric) generalized hyperbolic distribution is another important special case of the VMN distribution. It arises when W follows a generalized inverse Gaussian (GIG) distribution, which includes the IG distribution as a special case. Thus, the above-mentioned t-distribution and its nested cases are also members of the symmetric generalized hyperbolic distribution.

5 Parameter Estimation for MMN, MVMN, and VMN Distributions

The MMN, MVMN, and VMN distributions can be conveniently expressed in hierarchical forms that facilitate maximum likelihood estimation of the parameters of the models via the Expectation–Maximization (EM) algorithm [15]. For example, utilizing the hierarchical representation of the MMN distribution (5), an EM algorithm can be derived for any specified distribution $h(w; \theta)$ of W. Technical details of the EM algorithm for the MMN distribution and some of its special cases can be found in [1].

The MVMN distribution admits a similar hierarchical representation as given by (13). For special cases of the MVMN distribution such as the GH, MVNBS, and MVNL distributions, explicit expressions for the implementation of the EM algorithm can be found in [13, 26, 29], respectively. Software implementations of some special/limiting cases of the MVMN distributions are available. For example, the GH distribution and some of its nested cases have been implemented in the R package ghyh [33].

From (18), a VMN distribution can be expressed in a hierarchical form given by $\boldsymbol{Y}|W = w \sim N_p(\boldsymbol{\mu}, w\boldsymbol{\Sigma})$ and (with a slight abuse of notation) $W \sim h(w; \boldsymbol{\theta})$. Technical details of the EM algorithm for this model can be found in many reports, for example, [23]. Software implementation for the VMN distribution and some of its special cases are readily available. For example, the recent R package nvmix [19] allows a user-specified quantile function for W, as well as special cases including inverse gamma and Pareto distributions; see also [20].

6 Conclusions

A concise description of three generalizations of the (multivariate) normal distribution has been presented. These families of flexible distributions arise by mixing the mean and/or weighting the variance matrix of a normal random variable. Two of these families, namely the variance mixture (VMN) and mean–variance mixture of normal (MVMN) distributions, have a relatively long history in the literature, whereas the third family (mean mixture of normal (MMN) distribution) was introduced more recently. Each of these families has their own merits and limits. We have presented their basic properties, some important special/limiting cases, and references for parameter estimation procedures.

Throughout this paper, we have focused on the case of the univariate scaling variable. Some recent proposals have considered adopting a multivariate scaling variable $\boldsymbol{W}$ (with a matrix scaling coefficient $\boldsymbol{\Delta}$). Reference [17] introduced the multiple scale distribution, based on the VMN model but with a multivariate $\boldsymbol{W}$ where its elements are independent of each other. In the case of VMN distributions, some characterizations of the (multivariate) skew-normal distribution allow for $\boldsymbol{W}$ to have a multivariate truncated normal distribution [7, 31]. More recently, a variant of the generalized hyperbolic distribution that was studied in [34] is an example of a MVMN distribution that adopts a multivariate $\boldsymbol{W}$.

Some further versions and/or generalizations of MVMN would be of interest for future investigation; for example, a scale mixture of MMN distributions (as suggested by [28]) and a MVMN distribution where different mixing variables can be used for the mean and variance of the normal random variable.

References

1. Abdi, M., Madadi, M., Balakrishnan, N., Jamalizadeh, A.: Family of mean-mixtures of multivariate normal distributions: properties, inference and assessment of multivariate skewness (2020). arXiv:200610018
2. Adcock, C., Azzalini, A.: A selective overview of skew-elliptical and related distributions and of their applications. Symmetry **12**, 118 (2020)
3. Allard, A., Soubeyrand, S.: Skew-normality for climatic data and dispersal models for plant epidemiology: when application fields drive spatial statistics. Spatial Stat. **1**, 50–64 (2012)

4. Andrews, D.F., Mallows, C.L.: Scale mixtures of normal distributions. J. R. Stat. Soc. Ser. B **36**, 99–102 (1974)
5. Arellano-Valle, R.B., Azzalini, A.: On the unification of families of skew-normal distributions. Scandinav. J. Stat. **33**, 561–574 (2006)
6. Arellano-Valle, R.B., Azzalini, A.: A formulation for continuous mixtures of multivariate normal distributions (2020). arXiv:200313076
7. Arellano-Valle, R.B., Genton, M.G.: On fundamental skew distributions. J. Multivar. Anal. **96**, 93–116 (2005)
8. Asparouhov, T., Muthén, B.: Structural equation models and mixture models with continuous non-normal skewed distributions. Struct. Equ. Model. (2015). https://doi.org/10.1080/10705511.2014.947375
9. Azzalini, A.: The skew-normal distribution and related multivariate families. Scandinavian J. Stat. **32**, 159–188 (2005)
10. Azzalini, A., Capitanio, A.: The Skew-Normal and Related Families. Cambridge University Press, Cambridge (2014)
11. Azzalini, A., Dalla Valle, A.: The multivariate skew-normal distribution. Biometrika **83**, 715–726 (1996)
12. Barndorff-Nielsen, O., Kent, J., Sørensen, M.: Normal variance-mean mixtures and z distributions. Int. Stat. Rev. **50**, 145–159 (1982)
13. Browne, R.P., McNicholas, P.D.: A mixture of generalized hyperbolic distributions. Can. J. Stat. **43**, 176–198 (2015)
14. Contreras-Reyes, J.E., Arellano-Valle, R.B.: Growth estimates of cardinalfish (epigonus crassicaudus) based on scale mixtures of skew-normal distributions. Fisher. Res. **147**, 137–144 (2013)
15. Dempster, A.P., Laird, N.M., Rubin, D.B.: Maximum likelihood from incomplete data via the EM algorithm. J. R. Stat. Soc. B **39**, 1–38 (1977)
16. Deng, X., Yao, J.: On the property of multivariate generalized hyperbolic distribution and the stein-type inequality. Commun. Stat. Theory and Methods **47**, 5346–5356 (2018)
17. Forbes, F., Wraith, D.: A new family of multivariate heavy-tailed distributions with variable marginal amounts of tailweight: application to robust clustering. Stat. Comput. **24**, 971–984 (2014)
18. Genton, M.G. (ed.): Skew-Elliptical Distributions and Their Applications: A Journey Beyond Normality. Chapman & Hall, CRC, Boca Raton, Florida (2004)
19. Hintz, E., Hofert, M., Lemieux, C.: Normal variance mixtures: Distribution, density and parameter estimation (2019). arXiv:191103017
20. Hofert, M., Hintz, E., Lemieux, C.: nvmix: Multivariate Normal Variance Mixtures (2020). http://cran.r-project.org/web/packages/nvmixe, R package version 0.0-4
21. Iversen, D.: The generalized hyperbolic model:estimation, financial derivatives, and risk measures. Master's thesis, Albert-Ludwigs-Universität Freiburg (1999)
22. Kim, J.H.T., Kim, S.Y.: Tail risk measures and risk allocation for the class of multivariate normal mean–variance mixture distributions. Insuran.: Math. Econ. **86**, 145–157 (2019)
23. Lange, K., Sinsheimer, J.S.: Normal/independent distributions and their applications in robust regression. J. Comput. Graph. Stat. **2**, 175–198 (1993)
24. Lee, S., McLachlan, G.: Scale mixture distribution. Wiley Stats Ref: Statistics Reference Online (WSR). p. 08201 (2019)
25. McNeil, A.J., Frey, R., Embrechts, P.: Quantitative Risk Management: Concepts Techniques and Tools. Princeton University Press, New Jersey, US (2005)
26. Naderi, M., Arabpour, A., Jamalizadeh, A.: Multivariate normal mean-variance mixture distribution based on Lindley distribution. Commun. Stat.-Simul. Comput. **47**, 1179–1192 (2018)
27. Naderi, M., Bekker, A., Arashi, M., Jamalizadeh, A.: A theoretical framework for landsat data modeling based on the matrix variate mean-mixture of normal model. PLOS ONE **15**(4), e0230,773 (2020)
28. Negarestani, H., Jamalizadeh, A., Shafiei, S., Balakrishnan, N.: Mean mixtures of normal distributions: properties, inference and application. Metrika **82**, 501–528 (2019)

29. Pourmousa, R., Jamalizadeh, A., Rezapour, M.: Multivariate normal mean variance mixture distribution based on Birnbaum Saunders distribution. J. Stat. Comput. Simul. **85**, 2736–2749 (2015)
30. Pyne, S., Hu, X., Wang, K., Rossin, E., Lin, T.I., Maier, L.M., Baecher-Allan, C., McLachlan, G.J., Tamayo, P., Hafler, D.A., De Jager, P.L., Mesirow, J.P.: Automated high-dimensional flow cytometric data analysis. Proc. Natl. Acad. Sci. USA **106**, 8519–8524 (2009)
31. Sahu, S.K., Dey, D.K., Branco, M.D.: A new class of multivariate skew distributions with applications to Bayesian regression models. Can. J. Stat. **31**, 129–150 (2003)
32. Soltyk, S., Gupta, R.: Application of the multivariate skew normal mixture model with the EM algorithm to Value-at-Risk. In: Chan, F., Marinova, D., Anderssen, R.S. (eds.) MODSIM 2011 (19th International Congress on Modelling and Simulation), pp. 1638–1644. Perth, Australia (2011)
33. Weibel, M., Luethi, D., Breymann, W.: ghyp: Generalized Hyperbolic Distribution and Its Special Cases (2020). http://cran.r-project.org/web/packages/ghyp, R package version 1.6.1
34. Wraith, D., Forbes, F.: Location and scale mixtures of Gaussians with flexible tail behaviour: properties, inference and application to multivariate clustering. Comput. Stat. Data Anal. **90**, 61–73 (2015)

Robust Depth-Based Inference in Elliptical Models

Stanislav Nagy and Jiří Dvořák

Abstract Elliptical models are the most important family of multivariate probability distributions. We explore the properties of these distributions with respect to their halfspace depth and their illumination. The densities of elliptically symmetric distributions are expressed only in terms of the depth, the illumination, and a univariate function that can be estimated from the data. These observations set the ground for robust and nonparametric inference for (nearly) elliptical models based on the use of depth and illumination.

Keywords Elliptical distribution · Halfspace depth · Illumination · Density estimation

1 Depth and Illumination in Statistical Analysis

The *(halfspace) depth* [2, 13] is a notion that allows extensions of nonparametric statistical inference to multivariate data. Recall that the depth of a point $x \in \mathbb{R}^d$ with respect to (w.r.t.) the distribution of a random vector $X \sim P \in \mathscr{P}\left(\mathbb{R}^d\right)$ is defined by

$$hD\,(x; P) = \inf_{u \in \mathbb{R}^d} \mathsf{P}\left(\langle X, u\rangle \leq \langle x, u\rangle\right).$$

Here, $\mathscr{P}\left(\mathbb{R}^d\right)$ is the set of all (Borel) probability measures on $\mathbb{R}^d$, and $(\Omega, \mathscr{A}, \mathsf{P})$ is the probability space on which all random elements are defined. The depth is a measure of centrality of x, as evaluated w.r.t. the mass of P—the higher the depth of x is, the more appropriate it is to use x as a location estimator for P. The upper level sets of the depth, given for $\delta > 0$ by $P_\delta = \left\{y \in \mathbb{R}^d : hD\,(y; P) \geq \delta\right\}$, are called

S. Nagy (✉) · J. Dvořák
Faculty of Mathematics and Physics, Charles University, Prague, Czech Republic
e-mail: nagy@karlin.mff.cuni.cz

J. Dvořák
e-mail: dvorak@karlin.mff.cuni.cz

S. Balzano et al. (eds.), *Statistical Learning and Modeling in Data Analysis*, Studies in Classification, Data Analysis, and Knowledge Organization,
https://doi.org/10.1007/978-3-030-69944-4_14

the *central regions* of P. Shapes of these regions provide information about the geometric properties of the distribution. The depth is known to possess many nice properties—it is affine invariant and quite robust near the center of the distribution [2], that is in non-empty regions P_δ with δ large. In the tails of P, or more generally, at points where $hD(\cdot; P)$ is low, the depth loses its fine robustness properties and may be hard to be estimated accurately [3].

In [9], the latter problems were addressed using the concept of the illumination, a notion complementary to the depth. Suppose that $\alpha > 0$ is given so that the interior of P_α is non-empty. The (α-) *illumination* of x w.r.t. P is defined as the ratio of the volume of the convex hull of x and the central region P_α, and the volume of P_α itself

$$\mathscr{I}(x; P) = \mathrm{vol}_d(\mathrm{co}(x, P_\alpha)) / \mathrm{vol}_d(P_\alpha).$$

It is argued in [9] that in the tails of the distribution, the illumination is a better indicator of centrality than the depth. In contrast to the depth, the illumination is an outlyingness function—for $x \in P_\alpha$, the illumination equals one; for x outside P_α, it is greater than one and increases as x moves further from the central region. For a more detailed account of illumination and examples, see also [10].

2 Elliptically Symmetric Distributions

We are concerned with the properties of the halfspace depth and the illumination when applied to elliptically symmetric distributions. Recall that the distribution of a random vector $X = (X_1, \ldots, X_d)^\mathsf{T} \sim P \in \mathscr{P}(\mathbb{R}^d)$ is *spherically symmetric* if for any orthogonal matrix $O \in \mathbb{R}^{d\times d}$ we have $OX \sim P$. If a spherically symmetric random vector possesses a density f, it takes the form $f(x) = h(x^\mathsf{T}x)$ for all $x \in \mathbb{R}^d$. The function $h\colon [0, \infty) \to [0, \infty)$ is called the *density generator* of P. Clearly, h characterizes P. It is well known that all univariate projections $u^\mathsf{T}X$ of a spherically symmetric random vector X, with u in the unit sphere $\mathbb{S}^{d-1} \subset \mathbb{R}^d$, have the same distribution [4, Theorem 2.4]. In particular, the cumulative distribution function (c.d.f.) $F\colon \mathbb{R} \to [0, 1]$ of the first marginal X_1 (or any other univariate projection) also characterizes P completely. We call F the *marginal c.d.f.* of X.

For a spherically symmetric random vector $Z = (Z_1, \ldots, Z_d)^\mathsf{T}$ with the marginal c.d.f. F and density generator h, a center $\mu \in \mathbb{R}^d$, and a non-singular[1] matrix $A \in \mathbb{R}^{d\times d}$, the random vector $X = AZ + \mu$ is said to have an *elliptically symmetric distribution* with location μ and shape A. The distribution of X is described by the triplet (μ, A, F), but actually it can be shown [4, Sect. 2.1] that it depends only on (μ, Σ, F) for $\Sigma = AA^\mathsf{T}$. We therefore write $P = EC(\mu, \Sigma, F)$. The representation (μ, Σ, F) is not unique—a positive multiple of Σ, and an appropriately transformed

[1]For singular or non-square matrices A, results analogous to those given in this note can be shown if we restrict to the affine subspace given by the support of X.

F may lead to the same distribution P. This ambiguity is avoided by imposing that the determinant $|\Sigma|$ of Σ is one; we shall assume this in what follows.

Because $|\Sigma| = 1$, there exists a unique real, symmetric, and positive definite square root matrix $\Sigma^{1/2}$ of Σ [7, Theorem 7.2.6] that satisfies $\Sigma^{1/2}\Sigma^{1/2} = \Sigma$. Denote the inverse of $\Sigma^{1/2}$ by $\Sigma^{-1/2}$. This allows us to write $Z = \Sigma^{-1/2}(X - \mu)$. The density of $X \sim P = EC(\mu, \Sigma, F)$ is equal to

$$f(x) = h\left(\mathsf{d}_\Sigma(x, \mu)^2\right) \quad \text{for } x \in \mathbb{R}^d, \tag{1}$$

for h the density generator of Z. The density depends on x through the (Σ-)Mahalanobis distance of x from μ defined by $\mathsf{d}_\Sigma(x, \mu) = \sqrt{(x - \mu)^\mathsf{T} \Sigma^{-1}(x - \mu)}$. Denote by $\mathscr{E}_{\mu,\Sigma} = \left\{x \in \mathbb{R}^d : \mathsf{d}_\Sigma(x, \mu) \le 1\right\}$ the (Σ-)Mahalanobis ellipsoid around μ. Our task is to express the density f of $P = EC(\mu, \Sigma, F)$ in terms of the depth and the illumination of P. This allows estimation of f in a robust and nonparametric way.

2.1 *Depth and Illumination of Elliptical Distributions*

Elliptically symmetric distributions are quite simple when it comes to their depth. Since hD is affine invariant [2], the depth of x w.r.t. $X \sim P = EC(\mu, \Sigma, F)$ equals the depth of $z = \Sigma^{-1/2}(x - \mu)$ w.r.t. the spherically symmetric $\Sigma^{-1/2}(X - \mu) = Z \sim Q \in \mathscr{P}\left(\mathbb{R}^d\right)$ with the same density generator h as X. For spherically symmetric distributions, it is well known that the depth takes the form

$$hD(z; Q) = F\left(-\sqrt{\left(\Sigma^{-1/2}(x - \mu)\right)^\mathsf{T} \Sigma^{-1/2}(x - \mu)}\right) = 1 - F\left(\mathsf{d}_\Sigma(x, \mu)\right). \tag{2}$$

In particular, the central regions P_δ take the shape of Mahalanobis ellipsoids of P.

Because P_α is an ellipsoid, and because of [9, Lemma 1], we know that the illumination $\mathscr{I}(\cdot; P)$ is also a known function of the Mahalanobis distance $\mathsf{d}_\Sigma(x, \mu)$. Consequently, by [9, formula (10)], we can write for any $x \notin P_\alpha$

$$\mathsf{d}_\Sigma(x, \mu) = F^{-1}(1 - \alpha)\, g_d^{-1}\left(\mathscr{I}(x; P)\right) \tag{3}$$

for $g_d \colon [1, \infty) \to [1, \infty)$ a continuous, strictly increasing function defined by $g_d(1) = 1$, and for $t > 1$ as a primitive function to

$$g_d'(t) = \frac{\Gamma\left(\frac{d}{2} + 1\right)}{\sqrt{\pi}\,\Gamma\left(\frac{d+1}{2}\right)} \frac{1}{d} \left(1 - \frac{1}{t^2}\right)^{(d-1)/2} \qquad \text{for } t \in (1, \infty).$$

By g_d^{-1} we mean the inverse function to g_d, and F^{-1} stands for the quantile function corresponding to F. For additional details about g_d^{-1} and its properties, we refer to [9, Appendix A].

3 Depth-Based Density Estimation

The correspondence between elliptically symmetric distributions and the halfspace depth described in the previous section is well documented in the literature [5, 6]. Nonetheless, these relations have never been made explicit, and they have not been used to estimate the density f based solely on the halfspace depth/illumination.

For $P = EC(\mu, \Sigma, F)$ with density (1), we can express f in terms of the depth of P. Indeed, from (2) and (3), we get, given that $|\Sigma| = 1$,

$$f(x) = \begin{cases} h\left(\left(F^{-1}(1 - hD(x; P))\right)^2\right) & \text{if } hD(x; P) \geq \alpha, \\ h\left(\left(F^{-1}(1-\alpha)\, g_d^{-1}(\mathscr{I}(x; P))\right)^2\right) & \text{if } hD(x; P) < \alpha. \end{cases}$$

Therefore, to write f by means of the depth and the illumination, it remains to find depth-based expressions for the marginal c.d.f. F and the density generator h. Contrary to the approaches based exclusively on the halfspace depth, the advantage of using the illumination is that robustness and better stability properties, especially in the tails of P, can be achieved. This will be seen in an application in Sect. 4.

3.1 The Density Generator as the Volume of P_δ

Functions F and h are in a one-to-one relationship, i.e., one can be explicitly computed from the other. A closed form expression for this relation can be found, for instance, in [4, Sect. 2.2.3]. Write $f_{X_1} = F'$ for the density of X_1, and set $h_1(x) = f_{X_1}(x^2)$ for $x > 0$. For almost every $x > 0$ and m a positive integer, define

$$\begin{aligned} h_{m-1}(x) &= \int_x^\infty (y - x)^{-1/2}\, h_m(y)\, \mathrm{d}y, \\ h_{m+2}(x) &= -\frac{1}{\pi} h_m'(x). \end{aligned} \tag{4}$$

Then $h(x) = h_d(x)$ for almost every $x > 0$. Perhaps more elegantly, h can be also written as a fractional derivative of h_1 [14, Sect. 7.4].

We proceed by writing h and F in terms of the depth $hD(\cdot; P)$. Because of (2),

$$P_\delta = \left\{x \in \mathbb{R}^d : \mathsf{d}_\Sigma(x, \mu) \leq -F^{-1}(\delta) = F^{-1}(1-\delta)\right\} = \mathscr{E}_{\mu, \Sigma\left(F^{-1}(1-\delta)\right)^2} \tag{5}$$

for $\delta \in (0, 1/2)$. First, we compute the volume of P_δ. Any ellipsoid $\mathscr{E}_{\mu,\Sigma}$ is an affine image of the unit ball $B^d = \left\{x \in \mathbb{R}^d : \|x\| \leq 1\right\}$ given by $\mathscr{E}_{\mu,\Sigma} = \Sigma^{1/2} B^d + \mu = \bigcup_{x \in B^d} \left\{\Sigma^{1/2} x + \mu\right\}$. It is also well known that $\mathrm{vol}_d\left(B^d\right) = \pi^{d/2} / \Gamma(d/2 + 1)$ for $\Gamma(\cdot)$ the gamma function. This means that $\mathrm{vol}_d\left(\mathscr{E}_{\mu,\Sigma}\right) = \sqrt{|\Sigma|}\pi^{d/2} / \Gamma(d/2 + 1)$, and in particular also

$$\mathrm{vol}_d\left(P_\delta\right) = \left(F^{-1}\left(1-\delta\right)\right)^d \frac{\pi^{d/2}}{\Gamma\left(d/2+1\right)}.$$

The marginal c.d.f. F can be expressed from the formula above by

$$F\left(\frac{\left(\mathrm{vol}_d\left(P_\delta\right)\Gamma\left(d/2+1\right)\right)^{1/d}}{\sqrt{\pi}}\right) = 1-\delta \quad \text{for } \delta \in (0, 1/2). \tag{6}$$

Having F written in terms of P_δ only, we can use formulas (4) to obtain an analogous result for the density generator h directly using a plug-in method.

3.2 *The Density Generator as the Probability of P_δ*

A practical implementation of the formula for h using (6) and (4) requires precise numerical differentiation and integration, or numerical fractional differentiation. Therefore, it might lead to numerically unstable results. Here, we use a different approach and express h using its relation with the density of the radial distribution of P [4, Theorem 2.9]. Recall that any spherically symmetric Z with marginal c.d.f. F can be expressed in the form[2] $Z \stackrel{d}{=} RU$ for R a non-negative random variable whose distribution is called the *radial distribution* of Z, and U an independent random vector with uniform distribution on $\mathbb{S}^{d-1}$ [4]. By formula (2.21) from [4], there exists a simple relation between the density f_{R^2} of R^2 and the density generator h

$$f_{R^2}(t) = \frac{\pi^{d/2}}{\Gamma\left(d/2\right)} t^{d/2-1} h(t) \quad \text{for } t \geq 0. \tag{7}$$

The density of R^2 can be expressed directly from the depth $hD\left(\cdot; P\right)$. We start from (5), and use [4, formula (2.34)] to write for $\delta \in (0, 1/2)$

$$\mathsf{P}\left(X \in P_\delta\right) = \mathsf{P}\left(\mathsf{d}_\Sigma\left(X, \mu\right) \leq F^{-1}\left(1-\delta\right)\right) = \mathsf{P}\left(R^2 \leq \left(F^{-1}\left(1-\delta\right)\right)^2\right). \tag{8}$$

Thus, the probability content of a central region of X is given by the distribution function of R^2, the link between them being F. Because F depends on the depth via (6) and the left-hand side of (8) is expressed in terms of the probabilities of P_δ only, this allows to express the distribution of R^2 based on the depth and the probabilities of the central regions. It turns out that this approach to the estimation of h requires only a single numerical differentiation in practice, unlike the direct approach using only the volume of P_δ and (4). Therefore, the procedure based on (8) is more stable and uses both the information about the volume and the probability content of P_δ to estimate the density of P.

---

[2] $\stackrel{d}{=}$ stands for "is equal in distribution".

3.3 Fisher Consistent Estimators of the Density

Let $X_1, \ldots, X_n$ be a random sample from P that corresponds to an empirical measure $P_n \in \mathscr{P}(\mathbb{R}^d)$. To estimate the density f, we start by estimating the marginal c.d.f. F. First, in (6), one replaces the volume of the population depth central region P_δ by the volume of the empirical depth central region $(P_n)_\delta$ of points whose depth $hD(\cdot; P_n)$ is at least δ. An estimator of F is obtained in the implicit form of (6) with P replaced by P_n. More specifically, let $v_n\colon (0, 1/2) \to [0, \infty)\colon \delta \mapsto \mathrm{vol}_d((P_n)_\delta)$. This is a non-increasing function, and for its (generalized) inverse function v_n^{-1}, we can express our estimator of F as

$$F_n(t) = 1 - v_n^{-1}\left(\frac{\left(t\sqrt{\pi}\right)^d}{\Gamma(d/2+1)}\right) \quad \text{for } t \geq 0. \tag{9}$$

With an estimator of F at hand, we can use any of the two approaches described above to estimate h by h_n. For the method based on formula (8), the left-hand side in (8) is estimated by its empirical counterpart $\sum_{i=1}^n \mathbb{I}[X_i \in (P_n)_\delta]/n$. Function F can be replaced by its estimator (9). We get an estimator of the c.d.f. of R^2

$$F_{n,R^2}(t) = \sum_{i=1}^n \mathbb{I}\left[X_i \in (P_n)_{1-F_n(\sqrt{t})}\right]/n \quad \text{for } t \geq 0. \tag{10}$$

To estimate h, we take a derivative of (an interpolant of) this function, and set by (7)

$$h_n(t) = F'_{n,R^2}(t)\frac{\Gamma(d/2)}{\pi^{d/2}}t^{1-d/2} \quad \text{for } t \geq 0. \tag{11}$$

Finally, we are prepared to estimate the density f from (1) using P_n by

$$f_n(x) = \begin{cases} h_n\left(\left(F_n^{-1}(1 - hD(x; P_n))\right)^2\right) & \text{if } hD(x; P_n) \geq \alpha, \\ h_n\left(\left(F_n^{-1}(1-\alpha)\, g_d^{-1}(\mathscr{I}(x; P_n))\right)^2\right) & \text{if } hD(x; P_n) < \alpha. \end{cases} \tag{12}$$

In the following theorem, we summarize the derivations we made in Sect. 3.

Theorem 1 *Let $P = EC(\mu, \Sigma, F) \in \mathscr{P}(\mathbb{R}^d)$ be such that $|\Sigma| = 1$, and let f be the density of P. Then, for any $\alpha \in [0, 1/2)$, the estimator (12) taken as a functional of the empirical measure $P_n \in \mathscr{P}(\mathbb{R}^d)$ of a random sample from P is a Fisher consistent estimator of the true density f.*

In its current form, our main result is interesting mostly from the theoretical point of view. Its potential applications include depth-based inference for elliptical models or construction of asymptotically optimal, nonparametric, and highly robust classification rules. For the special case of multivariate normal distributions, the latter application was explored in [9, Sect. 5.3] and [10, Sect. 3] with quite promising first results.

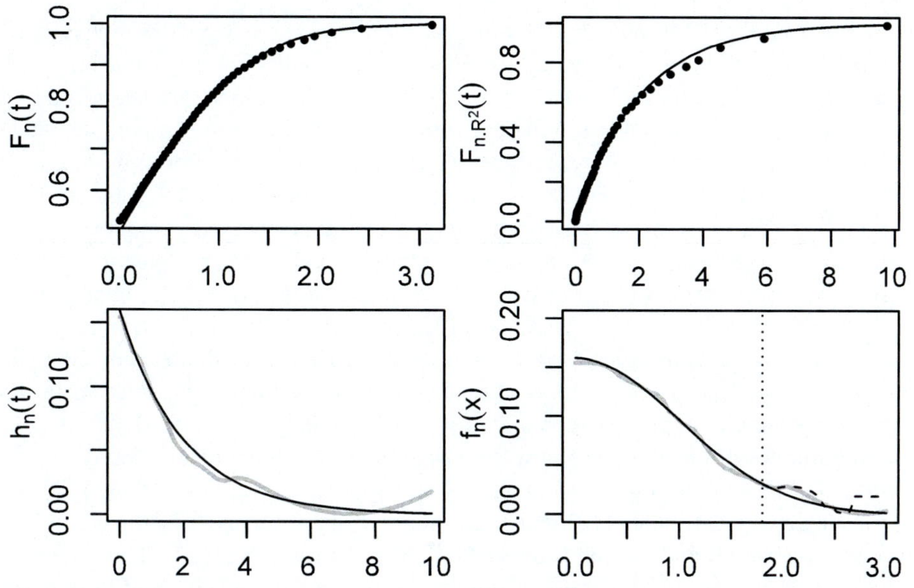

Fig. 1 Illustration of our nonparametric density estimation procedure for standard bivariate normal distribution. Solid black lines depict the true theoretical functions in all panels. Top left: estimate of the marginal c.d.f. F. Top right: estimate of the c.d.f. of R^2. Bottom left: estimate of the density generator h, shown as a thick gray curve. Bottom right: estimate of the density f using $\alpha = 0.05$, shown as a thick gray curve. The vertical dotted line indicates which estimated values are based solely on the depth (to the left from the vertical line) and which are based on a combination of the depth and illumination (to the right). For comparison, also the estimator using $\alpha = 0$, i.e., based only on the depth, is plotted by the black dashed line in the right part of the panel (in the left part of the panel, the estimator coincides with the original estimator and hence is not plotted)

4 Application

To illustrate the potential of the proposed nonparametric density estimation procedure, we employ it for P the standard bivariate normal distribution. In this case, the marginal c.d.f. F is the c.d.f. of the (univariate) standard normal distribution, R^2 has χ^2_2-distribution, and the density generator $h(t)$ is proportional to $\exp\{-t/2\}$, $t \geq 0$. We perform our computation in `R`, taking advantage of the packages `TukeyRegion` [1, 8] and `ddalpha` [11, 12]; the source code is available online.[3]

We generate $n = 500$ independent observations $X_i \sim P$ with an empirical measure P_n. A grid $\delta_1, \ldots, \delta_K$, used in further computations, is taken to be uniform in the interval $[\min_i hD(X_i; P_n), \max_i hD(X_i; P_n)]$, with $K = 51$. The volumes and the probability contents of the central regions $(P_n)_{\delta_j}$ are determined using the function `TukeyRegion`.

The values of the marginal c.d.f. F are estimated using (6) for $\delta = \delta_1, \ldots, \delta_K$, see the points in the top left panel of Fig. 1. Note the non-uniform sampling of points on

[3] `http://gems.karlin.mff.cuni.cz/software.php`.

the horizontal axis, implied by the uniform sampling of the δ_j values on the vertical axis. The distribution function of R^2 is estimated using (10) for a non-uniform grid of arguments $t_1, \ldots, t_K$ such that $\delta_j = 1 - F_n(\sqrt{t_j})$. The points corresponding to the estimate F_{n,R^2} are displayed in the top right panel of Fig. 1. To estimate the density generator h by (11), we approximate $F_{n,R^2}(t), t \geq 0$, by a smoothing spline and take the derivative of that spline, see the gray line in the bottom left panel of Fig. 1. Note that thanks to the spline approximation of the discretized estimate F_{n,R^2} it is possible to evaluate $h_n(t)$ at any $t \geq 0$. Finally, to estimate the density $f(x)$ for a given $x \in \mathbb{R}^2$, formula (12) is used. The required value of $F_n^{-1}(\cdot)$ is obtained from (6). For computation of g_d^{-1}, we refer to [9, Appendix A]. In our illustration, $f_n(x)$ is computed in a grid of points $x = (u, 0)^\mathsf{T}$ with $u \in [0, 3]$, see the bottom right panel of Fig. 1. For the cut-off value $\alpha = 0.05$, the estimated density is plotted in thick gray. For the choice $\alpha = 0$, the estimator is based solely on the depth and becomes unreliable for points with very low empirical depth, as illustrated by the dashed line plotted in the bottom right panel of Fig. 1.

Our density estimator (12) takes advantage of the robustness of the depth in the central region P_α and the robustness of the illumination outside P_α. This overcomes a major drawback of all depth-based procedures, namely their instability in regions of low depth. Hence α is an important tuning constant of our procedure, defining the largest central region where the depth estimates are still considered reliable. The amount of smoothing in the spline approximation of F_{n,R^2} is also specified by the user. Finally, the grid of values $\delta_1, \ldots, \delta_K$ can be chosen in a non-uniform way to obtain more detailed information in the tails of F_n and F_{n,R^2}. We leave these practical issues for further investigation.

The approach presented in this note is applicable in any dimension. Its only practical limitation arises from the need to compute the central regions $(P_n)_{\delta_j}$. This is a difficult task; the currently best-performing function `TukeyRegion` handles hundreds of observations in dimensions $d \leq 5$ without substantial difficulties [8].

Acknowledgements The authors thank two anonymous referees for their valuable comments on the original version of the manuscript. This work was supported by the grant 19-16097Y of the Czech Science Foundation and by the PRIMUS/17/SCI/3 project of Charles University.

References

1. Barber, C.B., Mozharovskyi P.: TukeyRegion: Tukey region and median (2019). R package version 0.1.2.1. https://CRAN.R-project.org/package=TukeyRegion
2. Donoho, D.L., Gasko, M.: Breakdown properties of location estimates based on halfspace depth and projected outlyingness. Ann. Stat. **20**(4), 1803–1827 (1992). https://doi.org/10.1214/aos/1176348890
3. Einmahl, J.H.J., Li, J., Liu, R.Y.: Bridging centrality and extremity: refining empirical data depth using extreme value statistics. Ann. Stat. **43**(6), 2738–2765 (2015). https://doi.org/10.1214/15-AOS1359

4. Fang, K.T., Kotz, S., Ng, K.W.: Symmetric multivariate and related distributions. Monographs on Statistics and Applied Probability, vol. 36. Chapman and Hall, Ltd., London (1990). https://doi.org/10.1007/978-1-4899-2937-2
5. Fraiman, R., Liu, R.Y., Meloche, J.: Multivariate density estimation by probing depth. In: L_1-statistical procedures and related topics (Neuchâtel, 1997). IMS Lecture Notes Monograph Series, Vol. 31, pp. 415–430. Institute of Mathematics Statistics, Hayward, CA (1997). https://doi.org/10.1214/lnms/1215454155
6. Ghosh, A.K., Chaudhuri, P.: On maximum depth and related classifiers. Scand. J. Statist. **32**(2), 327–350 (2005). https://doi.org/10.1111/j.1467-9469.2005.00423.x
7. Horn, R.A., Johnson, C.R.: Topics in Matrix Analysis. Cambridge University Press, Cambridge (1994). Corrected reprint of the 1991 original
8. Liu, X., Mosler, K., Mozharovskyi, P.: Fast computation of Tukey trimmed regions and median in dimension $p>2$. J. Comput. Graph. Stat. **28**(3), 682–697 (2019). https://doi.org/10.1080/10618600.2018.1546595
9. Nagy, S., Dvořák, J.: Illumination depth. J. Comput. Graph. Statist. **30**(1), 78–90 (2021). https://doi.org/10.1080/10618600.2020.1776717
10. Nagy, S., Dvořák, J.: Illumination in depth analysis. In: Porzio, G.C., Greselin, F., Balzano, S. (eds.) CLADAG 2019. Book of Short Papers, pp. 353–356. Università di Cassino e del Lazio Meridionale (2019)
11. Pokotylo O., Mozharovskyi P., Dyckerhoff R.: Depth and depth-based classification with R package ddalpha. J. Stat. Softw. **91**(5), 1–46 (2019). https://doi.org/10.18637/jss.v091.i05
12. Pokotylo, O., Mozharovskyi, P., Dyckerhoff, R., Nagy, S.: ddalpha: Depth-based classification and calculation of data depth (2017). R package version 1.3.1.1. https://CRAN.R-project.org/package=ddalpha
13. Tukey, J.W.: Mathematics and the picturing of data. In: Proceedings of the International Congress of Mathematicians (Vancouver, B. C., 1974), Vol. 2, pp. 523–531. Canad. Math. Congress, Montreal, Que. (1975)
14. Uchaikin, V.V., Zolotarev, V.M.: Chance and stability: Stable distributions and their applications. Modern Probability and Statistics. VSP, Utrecht (1999). https://doi.org/10.1515/9783110935974

Latent Class Analysis for the Derivation of Marketing Decisions: An Empirical Study for BEV Battery Manufacturers

Friederike Paetz

Abstract Currently, battery electric vehicles (BEVs) constitute prominent alternatives to vehicles with combustion motors. As the competition between BEV battery manufacturers increases, it is essential that they design batteries that perfectly meet the needs of BEV manufacturers. However, the needs of BEV manufacturers are actually derivatives of the needs of BEV customers. We, therefore, conducted an empirical discrete choice experiment with BEV customers in China and performed latent class analysis. We found substantial preference heterogeneity among BEV customers, which transfers into varying needs of BEV manufacturers w.r.t. batteries. Using these results, we determined the pricing and product design strategies for BEV battery manufacturers.

Keywords Latent class analysis · Discrete choice experiment · Marketing decisions

1 Motivation

Currently, battery electric vehicles (BEVs) constitute prominent alternatives to vehicles with combustion motors. As is documented, the demand for BEVs has tremendously increased in recent years. In 2014, approx. 750,000 BEVs were registered worldwide, and this number nearly quintupled to 3.2 million in 2017 [23]. Here, China has the most impressive growth rate for electromobility and has emerged as the most important market for BEVs [7].

Hence, it is not unexpected that Chinese BEV customers are the main focus in academic research. This attention is supported by the literature review of Ref. [13], which shows that recent studies have frequently investigated Chinese customers' preferences for BEV technologies. Often, these studies use discrete choice experiments (DCEs) to determine customer preferences for BEVs. Reference [9] conducted

F. Paetz (✉)
Department of Economics and Marketing, Clausthal Technical University, Julius-Albert-Str. 6, 38678 Clausthal-Zellerfeld, Germany
e-mail: friederike.paetz@tu-clausthal.de

S. Balzano et al. (eds.), *Statistical Learning and Modeling in Data Analysis*, Studies in Classification, Data Analysis, and Knowledge Organization,
https://doi.org/10.1007/978-3-030-69944-4_15

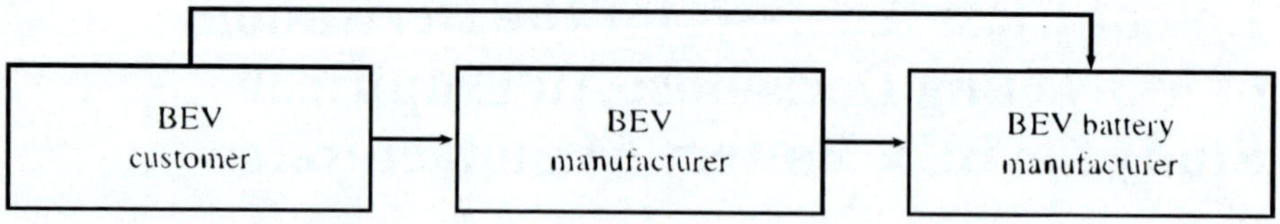

Fig. 1 Demand relationship

a DCE and estimated multinomial logit (MNL) models and mixed logit (MXL) models to compare the BEV preferences of Chinese and US customers.[1] Chinese customers showed a significantly higher relative willingness-to-pay (WTP) for BEV technology than did US customers. In addition, the authors investigated how subsidies influence the competitiveness of alternative technologies. To accomplish this task, they used a market simulation and found that Chinese customers' willingness to adopt BEVs was independent of subsidies. This result further replicates the result of an earlier study of Ref. [20], which found that the subsidy of waiving sales taxes does not have a significant effect on Chinese customers' willingness to buy BEVs. Reference [15] also focused on Chinese respondents and conducted a DCE. Like [5], they estimated MNL models and MXL models and explored the preference structures for certain BEV attributes/levels (driving range, charging time etc.). In addition, they compared the preferences of unlikely BEV customers and potential BEV customers and found that latter have a higher WTP for BEVs.

All studies show that Chinese respondents yield high preferences for BEVs and that they are willing to pay a price premium for BEVs in comparison to vehicles with combustion motors. Although the results from DCEs with stated choices are frequently criticized [27], they are supported in the focal BEV context by the observed tremendously increasing demand for BEVs in China in recent years [24].

With the increasing demand for BEVs, the demand for BEV batteries also derivatively increases. The battery is the core component of a BEV and is one of the main BEV cost drivers [26, p. 17]. Since the competition between BEV battery manufacturers is prevalent, BEV battery manufacturers have to accurately design their products and use sophisticated pricing strategies to gain a competitive advantage. A battery manufacturer could increase its competitive ability if its battery solves the problems associated with electromobility, namely, the driving range, (charging) infrastructure, and purchase price [5].

In sum, BEV customers evoke demand at BEV manufacturers, who, in turn, generate demand at BEV battery manufacturers. Hence, BEV customers indirectly generate demand for BEV batteries, as shown in Fig. 1. If we consider the demand relationship in Fig. 1, it becomes clear that the preferences of BEV customers influence the preferences of BEV manufacturers concerning BEV batteries. Hence, even in a business-to-business relationship, BEV battery manufacturers have to accommodate the BEV customers' preferences.

---

[1] For a concise guide to the MNL and MXL model please refer to [4].

This paper relies on the preference relationship explained above and explores the up-to-date preferences of Chinese customers in the electromobility research field. In particular, following the research methodology of recent studies, a DCE using BEVs was conducted to explore Chinese customers' preferences. However, we accommodate preference heterogeneity on a segment level and therefore differ from previous studies, which either do not account for preference heterogeneity at all (e.g., estimating MNL models) or capture heterogeneity on an individual level (e.g., estimating MXL models). The estimation of latent class multinomial logit (LC-MNL) models provides insights into preference structures of certain customer segments, i.e., submarkets that are homogeneous within their preferences. Segment-specific market contemplations are highly relevant for marketing practitioners to derive segment-specific marketing decisions. Here, the results of a LC-MNL model are used to draw inferences for the pricing and product strategies of BEV battery manufacturers. This paper is an advanced of Ref. [17]. This current version differs from Ref. [17] by providing new details regarding an elaborated literature review on Chinese preferences for BEVs and the demand relationships between BEV customers ans BEV battery manufacturers. In addition, a detailed description on the applied methodology, e.g., Latent Class Multinomial Logit Models, is provided.

The remainder of this contribution is as follows. In the next section, Sect. 2, we briefly introduce the LC-MNL model for the estimation of segment-specific preferences. Section 3 contains information on the empirical DCE. Here, the model selection procedure and the results of the selected segment solution are discussed. Furthermore, BEV customers' preferences are translated into inferences for pricing and product design strategies for BEV battery manufacturers. Finally, the results are concluded in Sect. 4.

2 Latent Class Multinomial Logit Model

The LC-MNL model is a widely used model in different application fields, e.g., marketing [11], and is available in various versions, e.g., using MNL parameterization with data collected over time [19]. The LC-MNL model relies on random utility theory [10, 14, 25]. Here, it is assumed, that a respondent j, where $j = 1, ..., J$, from a specific segment s, where $s = 1, ..., S$, chooses alternative m, where $m = 1, ..., M$, in a certain choice set t, where $t = 1, ..., T$, that provides the biggest utility U_{jstm} to that respondent. The utility vector $\mathbf{u}_{js} = (U_{js11}, ..., U_{jsTM})'$ contains the utilities of all alternatives in all choice sets and is considered to be a stochastic construct. It could be further described as the sum of a segment-specific deterministic part $\mathbf{v}_s$, $s = 1, ..., S$, and a stochastic component $\boldsymbol{\varepsilon}_j$: $\mathbf{u}_{js} = \mathbf{v}_s + \boldsymbol{\varepsilon}_j$. For the LC-MNL model, $\boldsymbol{\varepsilon}_j$, where $j = 1, ..., J$, is a random error term that is assumed to follow an i.i.d. Gumbel distribution. The deterministic term $\mathbf{v}_s = \mathbf{v}_{js}$ is the same for all respondents j in a certain segment s and could be further described as $\mathbf{v}_s = \mathbf{X}\boldsymbol{\beta}_s$. Here, $\mathbf{X}$ denotes the design matrix of the choice task, and $\boldsymbol{\beta}_s$ is the segment-specific part-worth utility vector of segment s.

For a fixed number of segments S, the researcher uses LC-MNL models to estimate both the segment-specific part-worth utility vectors $\boldsymbol{\beta}_s$ as well as the relative segment shares π_s, where $s = 1, ..., S$. The estimation is commonly performed using maximum likelihood estimation. Here, the log-likelihood can be maximized with iterative procedures such as the expectation-maximization algorithm [2, 12].

The log-likelihood is

$$LL = \prod_{j=1}^{J} \sum_{s=1}^{S} \pi_s \prod_{t=1}^{T} \prod_{m=1}^{M} P_{jstm}^{\delta_{jtm}}, \tag{1}$$

where P_{jstm} describes the choice probability of alternative m in choice set t by respondent j from segment s and δ_{jtm} is a binary variable, that equals one, if alternative m is chosen in choice set t by respondent j and zero otherwise. The choice probability is

$$P_{jstm} = \frac{exp(\mu \boldsymbol{x}'_{tm} \boldsymbol{\beta}_s)}{\sum_{r=1}^{R} exp(\mu \boldsymbol{x}'_{tr} \boldsymbol{\beta}_s)}, \tag{2}$$

where R denotes the number of alternatives in choice set t and $\mu > 0$ is a scale parameter. The design vector $\boldsymbol{x}'_{tm}$ of alternative m in choice set t is the corresponding row vector of X.

LC-MNL models are known for their fuzzy segment assignment, i.e., a respondent has a specific segment membership probability for each segment. Obviously, the better the separation between segments is, the more the posterior segment membership probabilities differ in favor of one specific segment [3]. For example, if two segments are well separated, the posterior segment membership of a respondent for one segment, e.g., segment 1, is likely to approximate 100%. However, if the two segments are very similar, i.e., less separated, then the posterior segment membership probability of a respondent is likely to approximate 50% for each of the two segments.

The extent of the heterogeneity within empirical (versus artificial) data sets is not known a priori. Therefore, LC-MNL models are estimated for several numbers of segments. To determine the appropriateness of certain segment solutions, different criteria have to be evaluated. For example, the model fit, the predictive validity, and the separation of segments provide valuable insights [18]. A popular measure to determine the model fit is the adjusted Bayesian information criterion (ABIC). The ABIC is calculated via [22]

$$ABIC = 2LL + ln\left(\frac{J+2}{24}\right)(S \cdot dim(\boldsymbol{\beta}_s)), \tag{3}$$

where LL denotes the log-likelihood value of the LC-MNL model and $dim(\boldsymbol{\beta}_s)$ describes the dimension of the part-worth utility vector $\boldsymbol{\beta}_s$. In comparison to the

Bayesian information criterion (BIC), that uses a penalty term of $ln(J)$, the ABIC uses penalty term of $ln(\frac{J+2}{24})$. Therefore, the ABIC circumvents the underestimation of the number of classes for small samples which is frequently reported for the BIC [16]. The mean posterior segment membership probabilities of respondents are frequently used as a measure of segments' separation and, therefore, model fit. Higher values represent a unique assignment of respondents to certain segments and a better separation of segments. These results provide a better model fit [3]. The predictive validity could be measured by the first choice (FC) hit rates in several holdout choice sets [18]. Obviously, the LC-MNL model of the segment solution with the highest FC hit rate performs the best.

3 Empirical Analysis

To gain information on customers' preferences for BEVs, we use the data of a DCE that was conducted in China. The DCE included 10 choice sets with three BEV alternatives and a no-purchase option. The BEVs were characterized by their driving range (150 km, 250 km, and 350 km), charging time (4 h, 6 h, and 8 h), purchase price (60, 000¥, 160, 000¥, and 260, 000¥), and car-body design (sedan, estate car, and SUV). All attributes were chosen in accordance with the attributes in the recent literature using DCEs to assess BEVs in China [15, 20]. In addition, the first three attributes cover the problems of electromobility. The attributes' levels conform to the most prevalent realizations of the top 20 best-selling BEV models in China in 2017 [6]. However, we did not incorporate Tesla because Tesla's BEVs are much more expensive and have a wider driving range and shorter charging time than all other top 20 BEVs in the Chinese market. The final sample includes 194 respondents.

The data of eight choice sets, which resulted in 1,552 observations, were used for the estimation of the LC-MNL models, and two holdout choice sets, which had 388 observations, were considered. The estimation was performed using the latent class module of the Sawtooth Software [21]. Furthermore, the effects-coding of all attributes and the part-worth utilities for all attribute levels were used. As a basis for the model selection, LC-MNL models for one to eight segments were estimated. We used the recommended convergence limit for the log-likelihood of 0.01 of Ref. [21]. Hence, the estimation procedure stops, if improvements in the log-likelihood between two iterations are less than 0.01. Furthermore, several criteria (e.g., the ABIC statistics, FC hit rates, and mean posterior segment membership probabilities) were calculated to determine the appropriateness of segment solutions.

Table 1 displays the results of the ABIC statistics, the FC hit rates, and the mean posterior segment membership probabilities (post. memb.).

The 6-segment solution achieves the best model fit, i.e., the lowest ABIC value, and the best predictive validity, i.e., the highest FC hit rate. The FC hit rate is 59% and therefore, the 6-segment solution performs more than two times better than chance (Each holdout choice set contained four alternatives, which results in a coin flip probability of a right prediction of $1/4 \cong 25\%$). In addition, the mean posterior

Table 1 Values of criteria for model selection (compare Ref. [17], p. 370)

Number of segments	ABIC	Mean post. memb. (%)	FC hit rates (%)
1	3,780	100	48
2	3,619	97	52
3	3,552	91	52
4	3,486	91	58
5	3,435	90	58
6	3,426	92	59
7	3,429	90	57
8	3,431	90	56

segment membership probability of the 6-segment solution (92%) is markedly higher than those of the 5- or 7-segment solutions (90%). This result argues for a less fuzzy segment assignment of respondents and, therefore, for well separated segments. All these criteria strongly argue for the selection of the 6-segment solution.

Table 2 illustrates the segment-specific part-worth utility estimates $\boldsymbol{\beta}_s$, the relative segment shares π_s, and the segment-specific relative attribute importances of the selected 6-segment solution. Obviously, the relative shares π_s of all six segments as well as the segment-specific attribute importances sum to 1 resp. 100%. In contrast, the segment-specific part-worth utility estimates of a certain attribute sum to 0, because we used effects-coding [1].

Table 2 shows that all segments are of meaningful size. Even the smallest segment, segment 4, with a relative segment size of 7%, contains 14 respondents and is therefore managerially worthwhile.

The interpretation of the segments may rely on the most preferred car-body design for the sake of simplicity. For example, segments with the highest preferences for SUVs could be interpreted as (potential) SUV customers, e.g., segment 1 and segment 6. SUV customers attach the highest importance to the driving range (39.16%) and purchase price attributes of a BEV (32.56%, segment 1), or the car-body design (60.28%, segment 6). Estate car customers (segment 3) attach the highest importance to the driving range (56.51%) and prefer higher prices, i.e., they view the price as a quality signal and favor the highest price of 260,000¥. Sedan customers either exclusively care about the car-body design (48.51%, segment 5) or about the attributes associated with the problems of electromobility (segment 2 and segment 4).

Obviously, well-separated preference structures between the considered BEV customer clusters exist. Those differences build a sound basis for customer-specific portfolio differentiations of BEV manufacturers and, therefore, derivatively for potential product and price differentiations of BEV battery manufacturers.

This approach is based on the characteristics of business-to-business (B2B) marketing. B2B buying behavior is based on derived demand, i.e., organizations (here: BEV manufacturers) buy products (here: BEV batteries) from other organizations (here: BEV battery manufacturers) that meet the needs of their customers [8, p. 193].

Table 2 Estimates of segment-specific shares, part-worth utilities, and attribute importances

rel. shares π_s	seg. 1	seg. 2	seg. 3	seg. 4	seg. 5	seg. 6
	0.361	0.162	0.103	0.070	0.184	0.120
Part-worth utilities β_s						
Driving range (in (km))						
150	−1.334	−2.044	−3.630	0.060	−0.224	−0.586
250	0.342	0.491	−0.162	−0.122	0.152	0.322
350	0.992	1.553	3.792	0.062	0.072	0.264
Charging time (in [h])						
4	0.221	0.621	1.716	−1.085	0.471	0.135
6	0.244	0.094	−0.353	0.193	−0.141	−0.158
8	−0.465	−0.715	−1.363	0.892	−0.330	0.023
Purchase price (in [¥])						
60,000	0.856	0.664	−1.157	−1.197	0.329	−0.561
160,000	0.222	0.072	0.102	0.660	0.223	0.168
260,000	−1.078	−0.736	1.055	0.537	−0.552	0.393
Car-body design						
estate car	0.175	−0.241	0.153	−0.225	−0.925	−0.142
sedan	−0.572	0.449	0.113	0.309	1.015	−1.564
SUV	0.397	−0.208	−0.266	−0.084	−0.090	1.706
Attribute importances (in [%])						
Driving range	39.16	51.23	56.51	4.06	9.40	16.73
Charging time	11.96	19.03	23.45	43.41	20.03	5.40
Purchase price	32.56	19.93	16.85	40.78	22.06	17.59
Car-body design	16.32	9.81	3.19	11.75	48.51	60.28

Hence, to derive inferences for a BEV battery manufacturer, we could ultimately rely on the preferences of different BEV customers (using Fig. 1). The batteries built for SUVs and estate cars need a high reservoir capacity because (potential) SUV customers and (potential) estate car customers prefer a wide driving range. Batteries for sedans must have a quick recharging time and a high reservoir capacity because the majority of (potential) sedan customers prefer a quick charging time and a wider driving range. (Potential) estate car customers view a higher price as a quality signal. Hence, the batteries for estate cars could be offered at higher prices and, therefore, may result in higher costs to fulfill customers' preferences for a wide driving range and a quick charging time.

For pricing and product design decisions, it could be inferred that even identically constructed batteries could be offered at higher prices to manufacturers of electric estate cars. Such a price differentiation arises from BEV manufacturers' opportunity to counteract higher battery costs by charging higher prices.

4 Conclusion

The rising demand of BEVs derivatively increases the demand for BEV batteries. Since the competition is strong, BEV battery manufacturers have to take the preferences of BEV manufacturers (and, therefore, the preferences of BEV customers) into account. We conducted a DCE using BEVs in China and estimated latent class multinomial logit models. Based on the results, inferences for pricing and product design strategies for BEV battery manufacturers were drawn. We found that the batteries built for SUVs, sedans, and estate cars need a high reservoir capacity. In addition, the batteries for sedans must have a quick recharging time. Furthermore, (potential) estate car customers viewed price as a quality signal. Therefore, the batteries for electric estate cars could be more expensive (and could be sold at higher prices) because BEV manufacturers could smoothly pass their higher costs onto their BEV customers.

Acknowledgements We thank Yundi Cheng for collecting the data.

References

1. Bech, M., Gyrd-Hansen, D.: Effects coding in discrete choice experiments. Health Econ. **14**, 1079–1083 (2005)
2. Dempster, A.P., Laird, N.M., Rubin, D.B.: Maximum likelihood from incomplete data via the EM algorithm. J. R. Stat. Soc. B. **39**, 1–38 (1977)
3. DeSarbo, Wayne S., Ramaswamy, V., Cohen, S. H.: Market segmentation with choice-based conjoint analysis. Mark. Lett. **6**, 137–147 (1995)
4. Elshiewy, O., Guhl, D., Boztug, Y.: Multinomial logit models in marketing from fundamentals to state of the art. Mark. ZFP. J. Res. Manag. **39**, 32–49 (2017)
5. Engel, H., Hensley, R., Knupfer, S., Sahdev, S.: Charging ahead: Electric-vehicle infrastructure demand. https://www.mckinsey.com/industries/automotive-and-assembly/our-insights/charging-ahead-electric-vehicle-infrastructure-demand. Cited 15 Nov 2019
6. EV Sales: China September 2017. http://ev-sales.blogspot.com/2017/10/china-september-2017.html. Cited 15 Nov 2019
7. Gong, H., Wang, M.Q., Wang, H.: New energy vehicles in China: policies, demonstration, and progress. Mitig. Adapt. Strateg. Glob. Change **18**, 207–228 (2013)
8. Grewald, R., Lilien, G.L., Bharadwaj, S., Jindal, P., Kayande, U., Lusch, R.F., Mantrala, M., Palmatier, R.W., Rindfleisch, A., Scheer, L.K., Spekman, R., Sridhar, S.: Business-to-business buying: challenges and opportunities. Cust. Need. Sol. **2**, 193–208 (2015)

9. Helveston, J.P., Liu, Y., McDonnell Feit, E., Fuchs, E., Klampfl, E., Michalek, J.J.: Will subsidies drive electric vehicle adoption? Measuring consumer preferences in the U.S. and China. Transp. Res. Part A **73**, 96–112 (2015)
10. Kamakura, W.A., Russell, G.: A probabilistic choice model for market segmentation and elasticity structure. J. Mark. Res. **26**, 379–390 (1989)
11. Leeflang, P., Wieringa, J.E., Bijmolt, T.H.A., Pauwels, K.H.: Advanced Techniques and Methods to Model Markets. Springer, Cham, Switzerland (2017)
12. Leisch, F.: FlexMix: a general framework for finite mixture models and latent class regression in R. J. Stat. Softw. **11**, 1–18 (2004)
13. Liao, F., Molin, E., van Wee, B.: Consumer preferences for electric vehicles: a literature review. Transp. Rev. **37**, 252–275 (2017)
14. McFadden, D.: Conditional logit analysis of qualitative choice behavior. In: Zarembka, P. (ed.) Frontiers in Econometrics, pp. 105–142. Academic, New York (1973)
15. Nie, Y., Wang, E., Guo, Q.: Examining Shanghai consumer preferences for electric vehicles and their attributes. In: Discussion Paper Series DP2017-21, Research Institute for Economics & Business Administration, Kobe University (2018). https://res.mdpi.com/d_attachment/sustainability/sustainability-10-02036/article_deploy/sustainability-10-02036.pdf. Cited 15 Nov 2019
16. Nylund, K.L., Asparouhov, T., Muthn, B.: Deciding on the number of classes in latent class analysis and growth mixture modeling. A Monte Carlo simulation study. Struct. Equ. Model. **14**, 535–569 (2007)
17. Paetz, F.: Price and product design strategies for manufacturers of electric vehicle batteries: inferences from latent class analysis. In: Porzio, Giovanni C., Greselin, Francesca, Balzano, Simona (eds.) CLADAG 2019 Book of Short Papers, pp. 369–372. Centro Editoriale di Ateneo Universitá di Cassino e del Lazio Meridionale, Cassino (2019)
18. Paetz, F., Hein, M., Kurz, P., Steiner, W.J.: Latent class conjoint choice models: A guide for model selection, estimation, validation, and interpretation of results. Mark. ZFP. J. Res. Manag. **41**, 3–20 (2019)
19. Pennoni, F., Paas, L., Bartolucci, F.: Causality patterns of a marketing campaign conducted over time: evidences from the latent Markov model. In: 49th Meeting of the Italian Statistical Society, Palermo 20-22 June 2018, 1–4 (2018)
20. Quian, L., Soopramanien, D.: Heterogeneous consumer preferences for alternative fuel cars in China. Transp. Res. Part D **16**, 607–613 (2018)
21. Sawtooth Software. The CBC Latent Class Technical Paper. Version 4 (2019) https://www.sawtoothsoftware.com/download/techpap/lctech.pdf? Cited 15 Jul 2020
22. Sclove, L.: Application of model-selection criteria to some problems in multivariate analysis. Psychometrika **52**, 333–343 (1987)
23. Statista. Worldwide number of battery electric vehicles in use from 2012 to 2017 https://www.statista.com/statistics/270603/worldwide-number-of-hybrid-and-electric-vehicles-since-2009/. Cited 15 Nov 2019
24. Statista. Estimated electric vehicles in use in selected countries as of 2018 https://www.statista.com/statistics/244292/number-of-electric-vehicles-by-country/. Cited 15 Nov 2019
25. Thurston, L.: A law of comparative judgment. Psychol. Rev. **34**, 273–286 (1927)
26. Tsiropoulos, I., Tarvydas, D., Lebedeva, N.: Li-ion batteries for mobility and stationary storage applications Scenarios for costs and market growth. EUR 29440 EN, Publications Office of the European Union, Luxembourg (2018)
27. Wardman, M.: A comparison of revealed preference and stated preference models of travel behaviour. J. Transp. Econ. Policy **22**, 71–91 (1988)

Small Area Estimation Diagnostics: The Case of the Fay–Herriot Model

Maria Chiara Pagliarella

Abstract Leverage and Cook's distance are some of the most important tools in influence analysis, where the main target is to identify observations that might determine the character of model estimates and predictors. In the small area estimation setup, applied statisticians are interested in tools to identify observations that might influence the variance component and the regression parameter estimates, the empirical best linear unbiased predictor and its mean squared error estimate. For this reason, this paper discusses the leverage matrix, the influence on the mean squared error of the empirical predictor, and a Cook's Distance of the empirical predictor for the Fay–Herriot model, when the area-random effect variance is estimated by the restricted maximum likelihood method. Further, the validity of this approach is illustrated by means of an application to poverty data.

Keywords Influence analysis · Leverage · Cook's distance · Poverty

1 Introduction

In the model-based approach to small area estimation, data is assumed to be generated according to a specific model and the whole inferential process depends on this assumption. Therefore, it is quite important to check if some data points or groups of cases are particularly influential on the analysis. For this reason, diagnostics tools are needed to ensure that model parameters are properly estimated.

In classical linear models, this examination has been traditionally carried out by residual analysis and detection of influential cases. Many articles and books deal with influential observations and outliers. Some of them are [3, 5, 21], and important papers have been written by Chatterjee and Hadi [8], and Cook [6, 7]. Two main types of influence analysis for linear models have been developed. Within the first, the calculation of leverage and standardized residuals plays a key role (the leverage

M. C. Pagliarella (✉)
Istituto Nazionale per l'Analisi delle Politiche Pubbliche (INAPP), Rome, Italy
e-mail: mc.pagliarella@inapp.org

S. Balzano et al. (eds.), *Statistical Learning and Modeling in Data Analysis*, Studies in Classification, Data Analysis, and Knowledge Organization,
https://doi.org/10.1007/978-3-030-69944-4_16

is the diagonal element of the hat matrix). The second one is based on measuring the effect on the estimates of deleting observations from the whole dataset, and it is called case deletion diagnostics. A third approach, less considered in applications, is based on the maximum curvature of log-likelihood displacement and it is called local influence (see [2, 7]).

In the context of mixed models, many contributions are available as well. Without this list being exhaustive, the following may be mentioned: Lesaffre and Verbeke [17] applied the local influence approach to linear mixed-effects models; Fung et al. [14] considered both case and subject deletion influence diagnostics for semi-parametric mixed models; Demidenko and Stukel [11] generalized common measures of influence for the fixed effects parameters of the linear mixed-effects models; Zewotir and Galpin [29] extended the ordinary linear regression influence diagnostics approach to linear mixed models; Nobre and Singer [22] covered a decomposition of the generalized leverage matrix for the linear mixed models; Pan et al. [23] proposed a case deletion approach to identify influential subjects and influential observations in linear mixed models.

A specific application of mixed models is small area estimation. Small area estimation refers to estimates over domains for which direct estimates are produced with unacceptably large standard errors due to the sample sizes available. Standard survey designs are typically carried out in order to achieve reliable estimates on planned domains (subpopulations) of the reference population. Direct estimates are those based only on the domain-specific sampling data. On the other hand, small area estimation produces indirect estimates for topic of interest on unplanned domains with too small or even zero sample sizes. Indirect estimators based on explicit linking models are called model-based estimators. They "borrow strength" by using values of the variables of interest from related small areas through supplementary information (auxiliary variables), such as data from other related areas or covariates from other sources.

Within this setting, case diagnostics requires special attention. Therefore, diagnostics for mixed models are an incomplete answer to diagnostics in small area estimation, because of the different population parameters of interest. This motivates our interest in diagnostics methods for area level linear mixed models appearing in small area estimation problems. In other words, the goal of small area estimation methods is to determine Empirical Best Linear Unbiased Predictor (EBLUP) for the mean or the total of the variable of interest and to minimize the Mean Squared Error (MSE) of the empirical predictor. Furthermore, case deletion diagnostics cannot be applied whenever there are few units for certain domain of interest.

However, while Battese, Harter, and Fuller [1] applied diagnostics methods for validating the small area estimation model, checking the normality of the error terms and the transformed residuals of the EBLUP, we found only a short note [20] on specific diagnostic measures for the Fay–Herriot model.

For these reasons, this paper has two main aims. On the one hand, it revises and makes it available to a larger audience the results in [20]. On the other, it shows the potential of such an approach by presenting an application of case diagnostics for the

Fay–Herriot model when the goal was to estimate poverty levels across small areas in Spain.

Fay–Herriot model is an area level linear mixed model, with random-area effects. It was first proposed by Fay and Herriot in 1979 [12] to estimate average per capita income in small places of the United States. Since then, the Fay–Herriot model has been widely used because of its flexibility in combining different sources of information with different error structures. It has been largely studied in small area estimation (e.g. [4, 9, 15, 18, 24]), and used to study poverty ([19, 25]) and other related socio-demographic variables [16].

The rest of the paper is organized as follows. Section 2 recalls the fundamentals of the area level Fay–Herriot model when we deal with Restricted Maximum Likelihood (REML) of the random-area effect variance estimator. Section 3 presents diagnostics for the Fay–Herriot model. More specifically Sect. 3.1 gives the leverage matrix on the fixed effects and the leverage matrix on the random-area effects; Sect. 3.2 shows the influence analysis on the first two terms of the estimated mean squared error of the EBLUP; Sect. 3.3 considers some case deletion diagnostics. Section 4 provides the application where diagnostics tools are tested on a model aiming at estimating small area poverty proportions in Spain, while Sect. 5 draws the conclusions. Lastly, an Appendix is provided with detailed formulas.

2 The Fay–Herriot Model

The Fay–Herriot model is a special case of a linear mixed model. We have

$$\widehat{y}_i = \mathbf{x}_i'\beta + b_i v_i + e_i, \quad v_i \overset{iid}{\sim} (0, \sigma_v^2), \quad e_i \overset{ind}{\sim} (0, \psi_i), \quad i = 1, ..., m \tag{1}$$

where the $\widehat{y}_i$'s are the direct estimates of the indicator of interest y for the i-th area, $\mathbf{x}_i$ is a vector containing the aggregated (population) values of p auxiliary variables with β regression coefficients, the random effects v_i and the sampling errors e_i are assumed to be independent with zero mean and known sampling variances ψ_i and unknown σ_v^2, respectively.

For our purposes, we rewrite the model in the general matrix form

$$\widehat{\mathbf{y}} = \mathbf{X}\beta + \mathbf{B}^{1/2}\mathbf{v} + \mathbf{e}, \tag{2}$$

where now $\mathbf{B} = \text{diag}(b_i^2)$ and the covariance matrix has a diagonal structure $\text{var}(\mathbf{y}) = \mathbf{V} = \text{diag}(V_i) = \text{diag}(\psi_i + \sigma_v^2 b_i^2)$. The vector of the Best Linear Unbiased Predictors (BLUPs) is given by

$$\widehat{\mathbf{y}}^H = \mathbf{X}\widehat{\beta} + \mathbf{B}^{1/2}\widehat{\mathbf{v}} = \mathbf{X}\widehat{\beta} + \mathbf{B}^{1/2}\sigma_v^2\mathbf{B}^{1/2}\mathbf{V}^{-1}(\widehat{\mathbf{y}} - \mathbf{X}\widehat{\beta}) = \mathbf{X}\widehat{\beta} + \Gamma(\widehat{\mathbf{y}} - \mathbf{X}\widehat{\beta}),$$

with $\Gamma = \text{diag}(\gamma_i) = \text{diag}(\sigma_v^2 b_i^2/(\psi_i + \sigma_v^2 b_i^2))$. The generalized least squares estimator of β is

$$\widehat{\beta} = \widehat{\beta}_{GLS} = (\mathbf{X}'\mathbf{V}^{-1}\mathbf{X})^{-1}(\mathbf{X}'\mathbf{V}^{-1}\widehat{\mathbf{y}}).$$

By using the relation

$$\mathbf{V}^{-1} = (\Psi + \sigma_v^2\mathbf{B})^{-1} = \Psi^{-1} - \Psi^{-1}(\Psi^{-1} + (\sigma_v^2)^{-1}\mathbf{B}^{-1})^{-1}\Psi^{-1} = \Psi^{-1}(\mathbf{I} - \Gamma)$$

where $\Psi = \text{diag}(\psi_i)$ and $\mathbf{I}$ is the identity matrix. Denoting with $\widehat{\mathbf{y}}^* = \Psi^{-1/2}\widehat{\mathbf{y}}$ and $\mathbf{X}^* = \Psi^{-1/2}\mathbf{X}$, for this estimator the result is

$$\begin{aligned}\widehat{\beta}_{GLS} &= (\mathbf{X}'\mathbf{V}^{-1}\mathbf{X})^{-1}(\mathbf{X}'\mathbf{V}^{-1}\widehat{\mathbf{y}})\\ &= (\mathbf{X}'\Psi^{-1}\mathbf{X} - \mathbf{X}'\Psi^{-1}\Gamma\mathbf{X})^{-1}(\mathbf{X}'\Psi^{-1}\widehat{\mathbf{y}} - \mathbf{X}'\Psi^{-1}\Gamma\widehat{\mathbf{y}})\\ &= (\mathbf{X}^{*\prime}\mathbf{X}^* - \mathbf{X}^{*\prime}\Gamma\mathbf{X}^*)^{-1}(\mathbf{X}^{*\prime}\widehat{\mathbf{y}}^* - \mathbf{X}^{*\prime}\Gamma\widehat{\mathbf{y}}^*).\end{aligned} \tag{3}$$

An Empirical Best Linear Unbiased Predictor (EBLUP) estimator is obtained from the BLUP by substituting suitable estimators of the variance and covariance parameters. Finally, the Restricted Maximum Likelihood (REML) estimator of σ_v^2 is (see [27] for more details)

$$\widehat{\sigma}^2_{v,REML} = \frac{a}{c^*}\frac{\sum(\widehat{y}_i^* - \overline{\widehat{y}}^*)^2 - (m-1)}{\sum(\widehat{y}_i^* - \overline{\widehat{y}}^*)^2}. \tag{4}$$

With reference to the error in the EBLUP estimator, Prasad and Rao in 1990 [24] gave an approximation to the mean squared error of the EBLUP under the Fay–Herriot model, which estimator includes three terms

$$mse(\hat{y}_i^H) = g_1(\hat{\sigma}_v^2) + g_2(\hat{\sigma}_v^2) + 2g_3(\hat{\sigma}_v^2). \tag{5}$$

It is worth noting that the terms g_2 and g_3, due to estimating β and σ_v, are of lower order than the leading term g_1.

The expressions (3), (4), and (5) will be used in next Section to derive the leverage matrix of the fixed and random effects, the influence on the MSE and a case deletion diagnostics for the empirical predictor.

3 Diagnostics for the Fay–Herriot Model

The main aim of a case diagnostics analysis is to identify observations or groups of observations that might determine the character of model estimates and predictors. In small area estimation, this means to identify the areas among the many that mostly affect the results of the estimates. In order to pursue that aim, after [20], we discuss three diagnostics measures: the leverage, the influential areas that affect the mean squared error estimates, and a Cook-type distance for the empirical predictor.

3.1 The Leverage Matrix

The aim is to investigate the influence of the domains (small areas) on the outcome of the analysis. We are therefore interested in the assessment of the effects of small perturbations in the data on the resulting BLUP estimates $\widehat{\mathbf{y}}^H$. For this reason, the leverage, that is the partial derivative of the predicted value with respect to the corresponding dependent variable, is considered here. In the framework of small area estimation under area level models, leverage is thus the partial derivative of the BLUP with reference to the corresponding direct estimator.

In order to obtain the leverage matrix of fixed and random effects, some useful results are provided below (more details are available within the Appendix). The leverage matrix for the traditional mixed model is given by definition as

$$L(\widehat{\beta}, \widehat{\mathbf{v}}) = \frac{\partial \widehat{\mathbf{y}}}{\partial \mathbf{y}}.$$

Assuming fixed $\mathbf{V}$, the leverage matrix $L(\widehat{\beta}, \widehat{\mathbf{v}})$ can be seen as sum of two components:

$$\begin{aligned} L(\widehat{\beta}, \widehat{\mathbf{v}}) &= L(\widehat{\beta}) + L(\widehat{\mathbf{v}}) \\ &= \mathbf{H}_1 + \mathbf{H}_2 \\ &= \mathbf{X}(\mathbf{X}'\widehat{\mathbf{V}}^{-1}\mathbf{X})^{-1}\mathbf{X}'\widehat{\mathbf{V}}^{-1} + \widehat{\sigma}^2_{v,REML}\mathbf{B}\widehat{\mathbf{P}}, \end{aligned}$$

where the first component is the hat matrix $\mathbf{H}_1 = \mathbf{X}(\mathbf{X}'\widehat{\mathbf{V}}^{-1}\mathbf{X})^{-1}\mathbf{X}'\widehat{\mathbf{V}}^{-1}$, also denominated generalized marginal leverage matrix, while the second component is given by $\mathbf{H}_2 = \widehat{\sigma}^2_{v,REML}\mathbf{B}\widehat{\mathbf{V}}^{-1}(\mathbf{I}_m - \mathbf{H}_1)$, the leverage matrix for the random component, with

$$\widehat{\mathbf{P}} = \widehat{\mathbf{V}}^{-1} - \widehat{\mathbf{V}}^{-1}\mathbf{X}(\mathbf{X}'\widehat{\mathbf{V}}^{-1}\mathbf{X})^{-1}\mathbf{X}'\widehat{\mathbf{V}}^{-1}.$$

Under model (2), it is appropriate to evaluate the effect of each area level direct estimate on the final predictor $\widehat{\mathbf{y}}^H$. The explicit form of the joint leverage matrix that we denote as L^* can be thus decomposed in terms of sampled observations as follows

$$L^*(\widehat{\beta}, \widehat{\mathbf{v}}) = \frac{\partial \widehat{\mathbf{y}}^H}{\partial \widehat{\mathbf{y}}} = \frac{\partial (\mathbf{X}\widehat{\beta})}{\partial \widehat{\mathbf{y}}} + \frac{\partial (\mathbf{B}^{1/2}\widehat{\mathbf{v}})}{\partial \widehat{\mathbf{y}}} = L^*(\widehat{\beta}) + L^*(\widehat{\mathbf{v}}).$$

Based on the derivative

$$\frac{\partial \mathbf{H}_1}{\partial \widehat{\mathbf{y}}} = \left[\frac{\partial (\mathbf{X}'\widehat{\mathbf{V}}^{-1}\mathbf{X})^{-1}}{\partial \widehat{\mathbf{y}}} (\mathbf{X}' \otimes \mathbf{X}') \right] (\widehat{\mathbf{V}}^{-1} \otimes \mathbf{I}_m) + \frac{\partial \widehat{\mathbf{V}}^{-1}}{\partial \widehat{\mathbf{y}}} (\mathbf{I}_m \otimes [\mathbf{X}(\mathbf{X}'\widehat{\mathbf{V}}^{-1}\mathbf{X})^{-1}\mathbf{X}']),$$

the leverage matrix for the fixed effects is given by [20]

$$\begin{aligned} L^*(\widehat{\beta}) &= \frac{\partial (\mathbf{X}\widehat{\beta})}{\partial \widehat{\mathbf{y}}} = \frac{\partial}{\partial \widehat{\mathbf{y}}} [\mathbf{X}(\mathbf{X}'\widehat{\mathbf{V}}^{-1}\mathbf{X})^{-1}\mathbf{X}'\widehat{\mathbf{V}}^{-1}\widehat{\mathbf{y}}] \\ &= \left(\frac{\partial \mathbf{H}_1}{\partial \widehat{\mathbf{y}}} \right) (\widehat{\mathbf{y}} \otimes \mathbf{I}_m) + \mathbf{H}_1' = \mathbf{H}_1^* + \mathbf{H}_1'. \end{aligned} \tag{6}$$

While the leverage associated with the estimated random effects is

$$\begin{aligned} L^*(\widehat{\mathbf{v}}) &= \frac{\partial (\mathbf{B}^{1/2}\widehat{\mathbf{v}})}{\partial \widehat{\mathbf{y}}} = \frac{\partial}{\partial \widehat{\mathbf{y}}} [\mathbf{B}^{1/2} \widehat{\sigma}^2_{v,REML} \mathbf{B}^{1/2} \widehat{\mathbf{V}}^{-1} (\widehat{\mathbf{y}} - \mathbf{X}\widehat{\beta})] \\ &= [(\mathbf{B} \otimes \sigma^2_{\partial})(\widehat{\mathbf{V}}^{-1} \otimes \mathbf{I}_m) + (\frac{\partial \widehat{\mathbf{V}}^{-1}}{\partial \widehat{\mathbf{y}}})(\mathbf{I}_m \otimes \widehat{\sigma}^2_{v,REML} \mathbf{B})][(\widehat{\mathbf{y}} - \mathbf{X}\widehat{\beta}) \otimes \mathbf{I}_m] \\ &+ [\mathbf{I}_m - L^*(\widehat{\beta})](\widehat{\sigma}^2_{v,REML} \mathbf{B}\widehat{\mathbf{V}}^{-1}). \end{aligned} \tag{7}$$

For the marginal leverage $\mathbf{H}_1$, as threshold value, it is suggested to use $2p/m$ (see [10]). Using $h^*_{1,ii}$ to indicate the diagonal elements of the matrix $\mathbf{H}_1^*$ (6) for the i-th area, and considering that $\text{tr}(\mathbf{H}_1') = p$, by analogy with [22], in our case influential observations that affect the fixed effects estimates can be verified comparing the quantity $(h^*_{1,ii} - \frac{1}{m}\text{tr}(\mathbf{H}_1^*))$ with $[L^*(\widehat{\beta})_{ii} - \frac{1}{m}\text{tr}(L^*(\widehat{\beta}))]$, more directly through the estimation of the model variance.

In practice, high-leverage observations are also identified by visual examination of the plot of the diagonal values of the leverage matrix. When we assess the potentially influential values, this is very useful in analyzing the contribution to the leverage of the single observation (small area) in estimating the model variance, with reference to the marginal leverage $\mathbf{H}_1$.

3.2 *Influence on the MSE of the EBLUP*

The final purpose in small area estimation is to determine EBLUP estimates for the mean or the total of the variable of interest and to minimize the mean squared error of the empirical predictor. Because of that, the influence of some small areas on the estimation of the MSE plays a central role in the analysis. Once an influential observation has been identified, it could therefore be removed by the researcher in order to improve the precision of the estimates.

With regard to the first component of the mean squared error estimate of the EBLUP, we have the following term as matrix form

$$\mathbf{G}_1 = \text{diag}(g_{1i}) = \Gamma\Psi \tag{8}$$

so that influential area estimates can be detected by the following derivative

$$\begin{aligned}\frac{\partial \mathbf{G}_1}{\partial \widehat{\mathbf{y}}} &= \frac{\partial}{\partial \widehat{\mathbf{y}}}(\Gamma\Psi) = \frac{\partial}{\partial \widehat{\mathbf{y}}}(\widehat{\sigma}^2_{v,REML}\mathbf{B}\widehat{\mathbf{V}}^{-1}\Psi) \\ &= (\mathbf{B} \otimes \sigma^2_{\partial})(\widehat{\mathbf{V}}^{-1}\Psi \otimes \mathbf{I}_m) + [\frac{\partial \widehat{\mathbf{V}}^{-1}}{\partial \widehat{\mathbf{y}}}(\Psi \otimes \mathbf{I}_m)](\mathbf{I}_m \otimes \widehat{\sigma}^2_{v,REML}\mathbf{B}). \end{aligned} \tag{9}$$

With reference to the second component, related to the variation of the fixed effects, that is

$$\mathbf{G}_2 = (\mathbf{I}_m - \Gamma)\mathbf{X}(\mathbf{X}'\widehat{\mathbf{V}}^{-1}\mathbf{X})^{-1}\mathbf{X}'(\mathbf{I}_m - \Gamma)(\mathbf{I}_m - \Gamma)\mathbf{U}(\mathbf{I}_m - \Gamma)],$$

after [20] we have the following influence measure

$$\frac{\partial \mathbf{G}_2}{\partial \widehat{\mathbf{y}}} = \frac{\partial (\mathbf{I}_m - \Gamma)}{\partial \widehat{\mathbf{y}}}([\mathbf{U}(\mathbf{I}_m - \Gamma)] \otimes \mathbf{I}_m) + \frac{\partial [\mathbf{U}(\mathbf{I}_m - \Gamma)]}{\partial \widehat{\mathbf{y}}}[\mathbf{I}_m \otimes (\mathbf{I}_m - \Gamma)], \tag{10}$$

where

$$\begin{aligned}\frac{\partial (\mathbf{I}_m - \Gamma)}{\partial \widehat{\mathbf{y}}} &= -(\mathbf{B} \otimes \sigma^2_{\partial})(\widehat{\mathbf{V}}^{-1} \otimes \mathbf{I}_m) + (\frac{\partial \widehat{\mathbf{V}}^{-1}}{\partial \widehat{\mathbf{y}}}\mathbf{I}_{m^2})(\mathbf{I}_m \otimes \widehat{\sigma}^2_{v,REML}\mathbf{B}), \\ \frac{\partial [\mathbf{U}(\mathbf{I}_m - \Gamma)]}{\partial \widehat{\mathbf{y}}} &= [\frac{\partial (\mathbf{X}'\widehat{\mathbf{V}}^{-1}\mathbf{X})^{-1}}{\partial \widehat{\mathbf{y}}}(\mathbf{X}' \otimes \mathbf{X}')][(\mathbf{I}_m - \Gamma) \otimes \mathbf{I}_m] + \frac{\partial (\mathbf{I}_m - \Gamma)}{\partial \widehat{\mathbf{y}}}(\mathbf{I}_m \otimes \mathbf{U}).\end{aligned}$$

These derivatives are important as they measure the increase (positive value) or the decrease (negative value) of the MSE of a small area, with reference to the direct estimate of another small area. For such an influential measure, no threshold values are available. Consequently, our suggestion is to first visualize the results for each area of interest (by column vector) from the $m \times m$ resulting matrix of (9) and (10) and then, for each column vector, investigate if there is any area showing particularly higher values.

3.3 Case Deletion Diagnostics and Cook's Distance

Here, we define Cook's distances for the REML estimate of $\hat{\sigma}^2_v$ and for the EBLUP $\hat{y}_i^H$.

After [20], Cook's distance for $\hat{\sigma}^2_v$ is given by

$$d^v_\ell = \frac{(\hat{\sigma}^2_v - \hat{\sigma}^2_{v(\ell)})^2}{\widehat{\text{var}}(\hat{\sigma}^2_v)},$$

where $\widehat{\text{var}}(\hat{\sigma}_v^2)$ is the asymptotic variance of $\hat{\sigma}_v^2$ that is obtained from the inverse of the REML Fisher information matrix, and the subscript $_{(\ell)}$ is used for those estimators that are calculated after deleting case ℓ.

The proposed Cook-type distance for the EBLUP $\hat{y}_i^H$ is

$$d_\ell^{eblup} = \frac{(\hat{y}_i^H - \hat{y}_{i(\ell)}^H)^2}{mse(\hat{y}_i^H)}$$

where $\hat{y}_{i(\ell)}^H$ is the EBLUP with case ℓ deleted and $mse(\hat{y}_i^H)$ is the Prasad-Rao [24] MSE estimator.

Cook's distance assesses the effects of a global change by removing an entire data point. It follows that large values of d_ℓ^{eblup} will point out that the corresponding area may affect the EBLUP estimate of the related deleted area.

4 An Application to Poverty Data

In order to design and implement poverty reduction policies and funding programs, there has been an increasing demand for poverty and living condition estimates at aggregate and local levels. Within such a framework, case diagnostics may have a very important role.

For this reason, this Section illustrates how the diagnostic tools introduced before can be exploited within a real data analysis performed on the official Spanish Living Condition Survey of the European Statistics on Income and Living Conditions (EU-SILC). The latter is a cross-sectional and longitudinal sample survey, coordinated by Eurostat, based on data from the European Union member states. It provides data on income, poverty, social exclusion, and living conditions in the European Union.

The analysis aims at estimating poverty levels in small area domains by the use of a Fay–Herriot area level model (1).

The dataset refers to the years 2004–2006 and contains 104 observations (areas in our context) obtained by crossing 52 Spanish provinces with 2 sex (men and women). The target variable is the direct estimate of the poverty indicator proposed by Foster et al. [13] (poverty incidence or proportion) at domain level (province × sex). Estimates of the domain means are used as responses in the area level model. The considered auxiliary variables are the known domain means of the category indicators of the following variables: age, education, citizenship, and labor. Finally, only 3 statistically significant variables that have a relevant meaning in a socio-economic sense are selected. They are age group 50–65, secondary education completed, and unemployment condition. The analysis was conducted with the open-source software R.

As discussed in Sect. 3.1, we start our influential analysis by computing the leverage values. We estimated the REML random-area effect variance, and we calculated

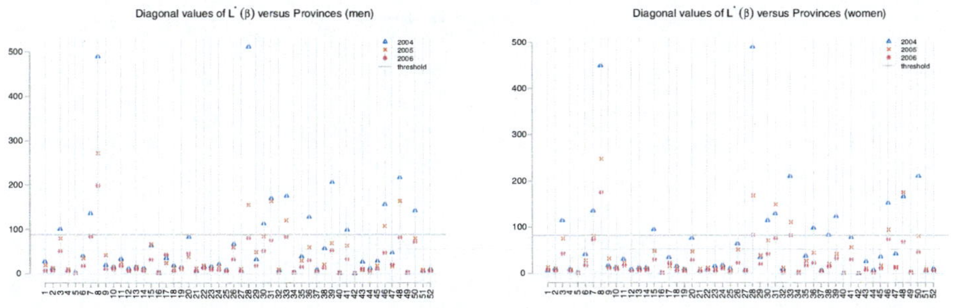

Fig. 1 The diagonal values of the leverage matrix for the fixed effects $L^*(\widehat{\beta})$ plotted for all the Spanish provinces, separately for men (left) and women (right)

the derivatives for the construction of the leverage matrix of the fixed effects as described in Eq. (6).

Results referring to the diagonal values of the leverage matrix for the fixed effects $L^*(\widehat{\beta})$ are presented in the scatter plot appearing in Fig. 1 for men and women, compared with a critical value of $2\left(\sum_{i=1}^{m} L^*_{ii}(\widehat{\beta})/m\right)$. By looking at the magnitude of the leverage values we conclude that the highest influential values are the provinces of Barcelona (8) and Madrid (28), for both sex categories, where the number in brackets indicates the corresponding numerical label on the abscissa axis of each plot. Differences also appear when focusing on to the three years taken into account. The leverage values tend to decrease from 2004 to 2006.

The relation between $L^*(\widehat{\beta})$ and the direct estimates of the poverty proportions along the three years 2004–2006 is shown in Fig. 2. The plot shows that the direct estimates with lower values generally correspond to higher level of leverage. Therefore, lower direct estimates of poverty proportion can be considered to be more influential than the higher ones. The values which stand out as the most influential are again the provinces of Barcelona (8) and Madrid (28).

Results of the influence analysis on the MSE of the EBLUP estimates, in particular that of the calculations of the derivative of $\mathbf{G}_1$ induced by the direct estimates (Eq. (9)), are illustrated in Fig. 3. By way of an example, according to their sample sizes (respectively small, large, and medium), the results for three provinces are presented: Alicante, Barcelona and Granada, for men and women, respectively, and for the year 2004. Similar results were also obtained for the years 2005 and 2006: for the sake of brevity they are not reported here. In this case, we observe a difference between men and women among the cases that are more influent on the MSE of the poverty level estimates. Among the three selected provinces, the most influential province for men is Granada, while for women it is Alicante. They are highlighted in the plots with dotted and dashed lines, respectively, which consistently dominate the others.

The derivative of $\mathbf{G}_1$ captures the influence that each small area can have on each other in terms of the power for increasing or decreasing the MSE of the EBLUP. The province of Granada suffers thus the increase of its MSE by the provinces of Badajoz (6), Barcelona (8), Madrid (28), Murcia (30), Sevilla (41), and Valencia

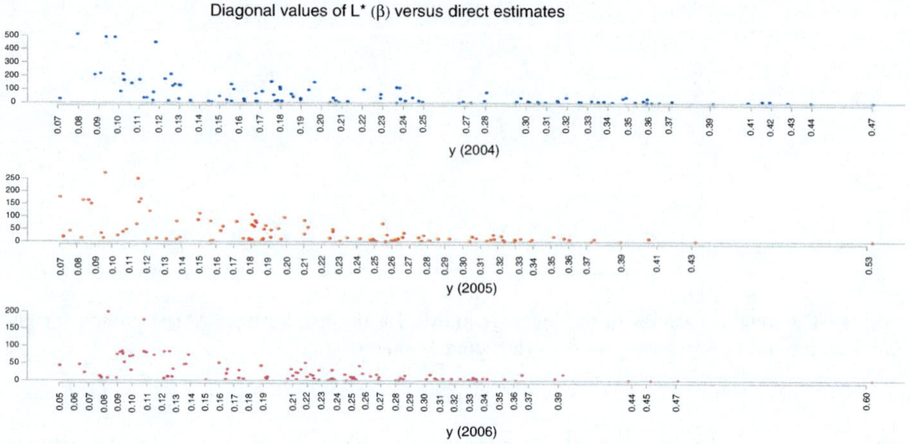

Fig. 2 The diagonal values of the leverage matrix for the fixed effects $L^*(\widehat{\beta})$ plotted against all the Spanish provinces direct estimates

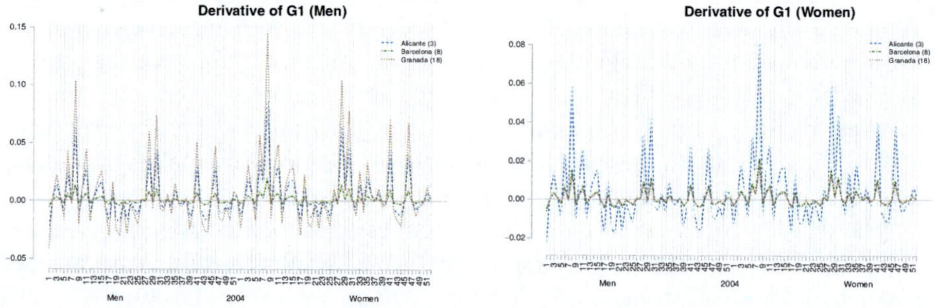

Fig. 3 Derivative of $\mathbf{G}_1$ for the provinces of Alicante (dashed line), Barcelona (dotdashed line), and Granada (dotted line) plotted against all the Spanish provinces in the year 2004, separately for men (left) and women (right)

(46). On the other side, the decrease of its MSE is caused by the provinces of Alava (1) and Gerona (17) (Fig. 3, left). As for the women (Fig. 3, right), the first part of the MSE of Alicante is affected by a positive influence by the same provinces that affect the plot of the men: Badajoz (6), Barcelona (8), Madrid (28), Murcia (30), Sevilla (41), and Valencia (46); the decrease instead is due to Alava (1), Gerona (17), Guipuzcoa (20), and Teruel (44).

Cook's Distance for the EBLUP as calculated in Sect. 3.3 are presented in Fig. 4. As done before, only the results of three provinces are presented: Alicante, Barcelona and Granada, for men and women, respectively, and for the year 2004. In the graph, three lines of Cook's distance are reported, which correspond to the deleted provinces of Alicante, Barcelona, and Granada. The peaks of Cook's distance represent the most influential values on the EBLUP estimates of the related deleted provinces. In particular, when looking at the results for men (Fig. 4, left), the lines of Alicante and

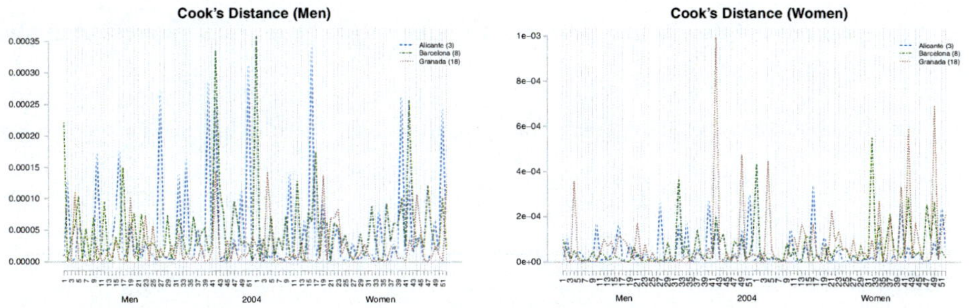

Fig. 4 Cook's Distance for the EBLUP $\hat{y}_i^H$ performed deleting the provinces of Alicante (dashed line), Barcelona (dotdashed line), and Granada (dotted line) plotted against all the Spanish provinces in the year 2004, separately for men (left) and women (right)

Barcelona show more peaks, and the deletion of these provinces shows that Alava (1), Cuenca (16), and Soria (42) are the provinces more influential for them. For women (Fig. 4, right), it shows that removing the province of Granada produces a high value of Cook's distance in correspondence with the province of Soria (42).

5 Concluding Remarks

A review of recent developments on diagnostic tools for the Fay–Herriot small area model when dealing with the restricted maximum likelihood estimate is proposed. Detailed formulae on fixed and random effects leverage matrices are reached in case of fixed $\mathbf{V}$. Tools for an influence analysis on the Mean Squared Error (MSE) of the Empirical Best Linear Unbiased Predictor (EBLUP) and a Cook's distance for the empirical predictor are considered.

The problem of the leverage of observed values on predicted values by the EBLUP was observed when we consider that, even though we make use of convenient estimates, the latter depends on the same influential values. Therefore, the leverage matrix of the model can be affected by influential observations through the estimates of the model variance. On the other hand, influence analysis on MSE estimates is based on $m \times m^2$-order matrices, which can be very useful in assessing the contribution of single observations (the small area direct estimates) in the evaluation of the MSE of all areas.

An application to real data is offered to the reader. The case of the estimation of poverty proportions for the Spanish provinces is exploited to illustrate the benefits of using specific diagnostic tools in the context of small area estimation. This methodology is useful because once the influential areas have been identified through visual examination, the researcher can eliminate them to improve the accuracy of the estimates.

Results in this paper are intended to be extended by the author to some other small area models, in particular to models that borrow strength from time or spatial correlations. It is thought that this research line might be of great interest to applied statisticians.

Acknowledgements The author thanks the Spanish National Statistics Institute for supplying the data.

Appendix

Details on how Eqs. (6), (7), (9) and (10) are derived are provided below.

For Eq. (6), let us first define the matrix A as

$$A = \sum(\widehat{y}_i^* - \widehat{\bar{y}})^2 = \widehat{\mathbf{y}}^{*\prime}\widehat{\mathbf{y}}^* - \frac{1}{m}(\widehat{\mathbf{y}}^{*\prime}\mathbf{1}_m)^2$$

where $\mathbf{1}_m$ denote the unitary vector all of whose components are unity. Following [20], we have then

$$\frac{\partial A}{\partial \widehat{\mathbf{y}}} = \frac{\partial}{\partial \widehat{\mathbf{y}}}\sum(\widehat{y}_i^* - \widehat{\bar{y}})^2 = \frac{\partial}{\partial \widehat{\mathbf{y}}}\left[\widehat{\mathbf{y}}^{*\prime}\widehat{\mathbf{y}}^* - \frac{1}{m}(\widehat{\mathbf{y}}^{*\prime}\mathbf{1}_m)^2\right]$$
$$= 2\widehat{\mathbf{y}}'\Psi^{-1} - \frac{2}{m}(\widehat{\mathbf{y}}'\mathbf{1}_m^{\Psi})(\mathbf{1}_m^{\Psi})', \text{ where } \mathbf{1}_m^{\Psi} = \Psi^{-1/2}\mathbf{1}_m.$$

For the Eqs. (7), (9), and (10), the derivative of the REML variance estimate is defined as follows:

$$\frac{\partial \widehat{\sigma}^2_{v,REML}}{\partial \widehat{\mathbf{y}}} = \frac{\partial}{\partial \widehat{\mathbf{y}}}\left[\frac{a}{c^*}\frac{A-(m-1)}{A}\right]$$
$$= \frac{a}{c^*}A^{-2}\left[2\widehat{\mathbf{y}}'\Psi^{-1} - \frac{2}{m}(\widehat{\mathbf{y}}'\mathbf{1}_m^{\Psi})(\mathbf{1}_m^{\Psi})'\right] = \sigma^2_{\partial}.$$

For the derivative of $\widehat{\mathbf{V}}$ and its inverse, appearing in Eqs. (7) and (9), we have the following:

$$\frac{\partial \widehat{\mathbf{V}}}{\partial \widehat{\mathbf{y}}} = \frac{\partial}{\partial \widehat{\mathbf{y}}}\text{diag}(\psi_i + \widehat{\sigma}^2_{v,REML}b_i^2)$$
$$= \frac{\partial}{\partial \widehat{\mathbf{y}}}(\Psi + \mathbf{B}^{1/2}\widehat{\sigma}^2_{v,REML}\mathbf{B}^{1/2}) = \mathbf{B}\otimes\sigma^2_{\partial},$$

$$\frac{\partial \widehat{\mathbf{V}}^{-1}}{\partial \widehat{\mathbf{y}}} = -(\widehat{\mathbf{V}}^{-1}\mathbf{B}\widehat{\mathbf{V}}^{-1})\otimes\sigma^2_{\partial}.$$

Finally, the derivative of $(\mathbf{X}'\widehat{\mathbf{V}}^{-1}\mathbf{X})^{-1}$ which refers to Eq. (10) is

$$\frac{\partial(\mathbf{X}'\widehat{\mathbf{V}}^{-1}\mathbf{X})^{-1}}{\partial\widehat{\mathbf{y}}} = -\frac{\partial(\mathbf{X}'\widehat{\mathbf{V}}^{-1}\mathbf{X})}{\partial\widehat{\mathbf{y}}}[(\mathbf{X}'\widehat{\mathbf{V}}^{-1}\mathbf{X}) \otimes (\mathbf{X}'\widehat{\mathbf{V}}^{-1}\mathbf{X})]$$

$$= [(\widehat{\mathbf{V}}^{-1}\mathbf{B}\widehat{\mathbf{V}}^{-1}) \otimes \sigma_{\partial}^{2}](\mathbf{X} \otimes \mathbf{X})[(\mathbf{X}'\widehat{\mathbf{V}}^{-1}\mathbf{X}) \otimes (\mathbf{X}'\widehat{\mathbf{V}}^{-1}\mathbf{X})].$$

References

1. Battese, G.E., Harter, R.M., Fuller, W.A.: An error component model for prediction of county crop areas using survey and satelite data. J. Am. Stat. Assoc. **83**, 28–36 (1988)
2. Beckman, R.J., Nachtsheim, C.J., Cook, R.D.: Diagnostics for mixed-model analysis of variance. Technometrics **29**, 413–426 (1987)
3. Belsley, D.A., Kuh, E., Welsh, R.E.: Regression Diagnostics: Identifying Influential Data and Sources of Collinearity. Wiley, Hoboken (1980)
4. Benavent, R., Morales, D.: Multivariate Fay? Herriot models for small area estimation. Comput. Stat. Data Anal. **94**, 372–390 (2016)
5. Cook, R.D., Weisberg, S.: Residuals and Influence in Regression. Chapman and Hall, New York (1982)
6. Cook, R.D.: Detection of influential observations in linear regression. Technometrics **19**, 15–18 (1977)
7. Cook, R.D.: Assessment of local influence. J. R. Stat. Soc. B **48**, 133–169 (1986)
8. Chatterjee, S., Hadi, A.S.: Influential observations, high leverage points, and outliers in linear regression. Stat. Sci. **1**, 379–416 (1986)
9. Datta, S., Lahiri, P.: A unified measure of uncertainty of estimated best linear unbiased predictors in small area estimation problems. Stat. Sinica **10**, 613–627 (2000)
10. Demidenko, E.: Mixed Models: Theory and Applications. Wiley, Hoboken (2004)
11. Demidenko, E., Stukel, T.A.: Influence analysis for linear mixed-effects models. Stat. Med. **24**, 893–909 (2005)
12. Fay, R.E., Herriot, R.A.: Estimates of income for small places: an application of James-Sein procedures to census data. J. Am. Stat. Assoc. **74**, 269–277 (1979)
13. Foster, J., Greer, J., Thorbecke, E.: A class of decomposable poverty measures. Econometrica **52**, 761–766 (1984)
14. Fung, W.K., Zhu, Z.Y., Wei, B.C., He, X.: Influence diagnostics and outliers tests for semiparametric mixed models. J. R. Stat. Soc. B **64**, 565–579 (2002)
15. Gonzalez-Manteiga, W., Lombardia, M.J., Molina, I., Morales, D., Santamaria, L.: Small area estimation under Fay?Herriot models with nonparametric estimation of heteroscedasticity. Stat. Model. **10**(2), 215–239 (2010)
16. Herrador, M., Esteban, M.D., Hobza, T., Morales, D.: A Fay?Herriot model with different random effect variances. Commun. Stat. Theory Methods **40**(5), 785–797 (2011)
17. Lesaffre, E., Verbeke, G.: Local Influence in Linear Mixed Models. Biometrics **54**(2), 570–582 (1998)
18. Marhuenda, Y., Molina, I., Morales, D.: Small area estimation with spatio-temporal Fay? Herriot models. Comput. Stat. Data Anal. **58**, 308–325 (2013)
19. Molina, I., Rao, J.N.K.: Small area estimation of poverty indicators. Can. J. Stat. **38**, 369–385 (2010)
20. Morales D., Pagliarella M.C., Salvatore R.: Influence analysis in small area estimation. In: Conference Proceedings of the 45th Meeting of the Italian Statistical Society. Cleup Editore, Padova (2010)

21. Neter, J., Kutner, M.H., Nachtsheim, C.J., Wasserman, W.: Applied Linear Statistical Models. IRWIN, Chicago (1990)
22. Nobre, J.S., Singer, M.: Leverage analysis for linear mixed models. J. Appl. Stat. **38**(5), 1063–1072 (2011)
23. Pan, J., Fei, Y., Foster, P.: Case-Deletion Diagnostics for Linear Mixed Models. Technometrics **56**(3), 269–281 (2014)
24. Prasad, N.G.N., Rao, J.N.K.: The estimation of the mean squared error of small-area estimators. J. Am. Stat. Assoc. **85**, 163–171 (1990)
25. Pratesi, M.: Analysis of Poverty Data by Small Area Estimation. Wiley, New York (2016)
26. Pregibon, D.: Logistic Regression Diagnostics. Ann. Stat. **4**, 705–724 (1981)
27. Rao, J.N.K.: Small Area Estimation. Wiley, Hoboken NJ (2003)
28. Singh, B., Shukla, G., Kundu, D.: Spatio-temporal models in small area estimation. Surv. Methodol. **31**, 183–195 (2005)
29. Zewotir, T., Galpin, J.S.: Influence diagnostics for linear mixed models. J. Data Sci. **3**, 153–177 (2005)

A Comparison Between Methods to Cluster Mixed-Type Data: Gaussian Mixtures Versus Gower Distance

Monia Ranalli and Roberto Rocci

Abstract In this paper, we compare through a simulation study two approaches to cluster mixed-type data, where some variables are continuous and some others ordinal. The first is model-based, according to which the variables are assumed to follow a Gaussian mixture model, where, as regards the ordinal variables, it is only partially observed. In order to overcome computational issues, the parameter estimation is carried out through an EM-like algorithm maximizing a composite log-likelihood based on low-dimensional margins. In the second approach, the Gower distance matrix is computed, then the PAM algorithm is used for clustering.

Keywords Mixture models · Composite likelihood · EM algorithm · Mixed-type data · Gower's distance · PAM algorithm

1 Introduction

The aim of cluster analysis is to partition the data into meaningful homogeneous groups which should differ considerably from each other. The problem is made more difficult by the presence of mixed-type data: ordinal and continuous variables. In order to find a solution, mainly two different approaches exist, based on a model describing the data generation process or a distance able to capture the dissimilarity between two entities. Before to summarize the main features of the two approaches, let us specify that when we use the word categorical data, we are still referring to the ordinal variables. Following the definition given in [1], ordinal variables are categorical variables with ordered categories.

As regards the model-based approach, the literature on clustering for continuous data is rich and wide; the most commonly clustering model-based used is the finite

M. Ranalli (✉) · R. Rocci
Department of Statistics, Sapienza University of Rome, Rome, Italy
e-mail: monia.ranalli@uniroma1.it

R. Rocci
e-mail: roberto.rocci@uniroma1.it

S. Balzano et al. (eds.), *Statistical Learning and Modeling in Data Analysis*, Studies in Classification, Data Analysis, and Knowledge Organization, https://doi.org/10.1007/978-3-030-69944-4_17

mixture of Gaussians [17]). Differently, that one developed for categorical data is still limited. In the Underlying Response Variable (URV), mainly developed in the SEM framework (see, e.g., [11, 14, 20] approach, the ordinal variables are seen as a discretization of continuous latent variables jointly distributed as a finite mixture (see [5, 16, 23]. However, this makes the maximum likelihood estimation rather complex because it requires the computation of many high-dimensional integrals. The problem is usually solved by approximating the likelihood function by a surrogate one. In this regard we mention some useful surrogate functions, such as the variational likelihood [7] or the composite likelihood [21, 23, 24]. The problem arises when we consider the joint distribution between continuous and ordinal variables. By assuming the local independence assumption, the issue can be easily solved by factorizing the joint density into the product of univariate marginals. However, this assumption is unrealistic and too restrictive.

Following the URV approach, [5, 23] proposed a model according to which the variables follow a Gaussian mixture model, where some variables, the ordinal ones, are only partially observed through their discretization. As a side note, at this stage, nominal variables cannot be included in the model, since there is no type of proximity among the unordered categories.

Besides these methods, there are others based on the Gower's distance [8]. This is computed as the average of partial dissimilarities across subjects (or entities), where the type of partial dissimilarity used depends on the specific type of the variable. To cluster the data then a k-medoids algorithm can be used (PAM algorithm, [13, 25]). However, these clustering methods are not the only ones existing in literature. Indeed there are many techniques for mixed-type data and many reviews. See, for example, [2, 6, 10]. Comparing clustering techniques is extremely useful and benchmarking in cluster analysis has been increasing. A good discussion on it can be found in [18].

The paper aims at exploring and comparing the behavior of the mixture model for mixed-type data with the distance-based methods, and some more naive approaches, according to which ordinal data are treated as metric.

The plan of the paper is as follows. In Sect. 2, we describe the model-based approach to cluster mixed-type data. The Gower distance method followed by the PAM algorithm is described in Sect. 3. In Sect. 4, we compare these clustering techniques through a simulation study. In the last section, some concluding remarks are pointed out.

2 The Model-Based Approach

Let $\mathbf{x} = [x_1, \ldots, x_O]'$ and $\mathbf{y}^{\bar{O}} = [y_{O+1}, \ldots, y_P]'$ be O ordinal and $\bar{O} = P - O$ continuous variables, respectively. The associated categories for each ordinal variable are denoted by $c_i = 1, 2, \ldots, C_i$ with $i = 1, 2, \ldots, O$.

Following the Underlying Response Variable (URV) approach, the ordinal variables $\mathbf{x}$ are considered as a categorization of a continuous multivariate latent variable

$\mathbf{y}^O = [y_1, \ldots, y_O]'$. The latent relationship between $\mathbf{x}$ and $\mathbf{y}^O$ is explained by the threshold model,

$$x_i = c_i \Leftrightarrow \gamma^{(i)}_{c_i-1} \leq y_i < \gamma^{(i)}_{c_i},$$

where $-\infty = \gamma^{(i)}_0 < \gamma^{(i)}_1 < \ldots < \gamma^{(i)}_{C_i-1} < \gamma^{(i)}_{C_i} = +\infty$ are the thresholds defining the C_i categories collected in a set $\boldsymbol{\Gamma}$. To accommodate both cluster structure and dependence within the groups, we assume that $\mathbf{y} = [\mathbf{y}^{O\prime}, \mathbf{y}^{\bar{O}\prime}]'$ follows a heteroscedastic Gaussian mixture, $f(\mathbf{y}) = \sum_{g=1}^{G} \tau_g \phi_P(\mathbf{y}; \boldsymbol{\mu}_g, \boldsymbol{\Sigma}_g)$, where the τ_g's are the mixing weights and $\phi_P(\mathbf{y}; \boldsymbol{\mu}_g, \boldsymbol{\Sigma}_g)$ is the density of a P-variate normal distribution with mean vector $\boldsymbol{\mu}_g$ and covariance matrix $\boldsymbol{\Sigma}_g$.

Let us set $\boldsymbol{\psi} = \{\tau_1, \ldots, \tau_G, \boldsymbol{\mu}_1, \ldots, \boldsymbol{\mu}_G, \boldsymbol{\Sigma}_1, \ldots, \boldsymbol{\Sigma}_G, \boldsymbol{\Gamma}\} \in \boldsymbol{\Psi}$, where $\boldsymbol{\Psi}$ is the parameter space. For a random i.i.d. sample of size N: $(\mathbf{x}_1, \mathbf{y}_1^{\bar{O}}), \ldots, (\mathbf{x}_N, \mathbf{y}_N^{\bar{O}})$, the log-likelihood is

$$\ell(\boldsymbol{\psi}) = \sum_{n=1}^{N} \log \left[\sum_{g=1}^{G} \tau_g \phi_{\bar{O}}(\mathbf{y}_n^{\bar{O}}; \boldsymbol{\mu}_g^{\bar{O}}, \boldsymbol{\Sigma}_g^{\bar{O}\bar{O}}) \pi_n \left(\boldsymbol{\mu}_{n;g}^{O|\bar{O}}, \boldsymbol{\Sigma}_g^{O|\bar{O}}, \boldsymbol{\Gamma} \right) \right], \qquad (1)$$

where with obvious notation

$$\pi_n \left(\boldsymbol{\mu}_{n;g}^{O|\bar{O}}, \boldsymbol{\Sigma}_g^{O|\bar{O}}, \boldsymbol{\Gamma} \right) = \int_{\gamma^{(1)}_{c_1-1}}^{\gamma^{(1)}_{c_1}} \cdots \int_{\gamma^{(O)}_{c_O-1}}^{\gamma^{(O)}_{c_O}} \phi_O(\mathbf{u}; \boldsymbol{\mu}_{n;g}^{O|\bar{O}}, \boldsymbol{\Sigma}_g^{O|\bar{O}}) d\mathbf{u}$$

$$\boldsymbol{\mu}_{n;g}^{O|\bar{O}} = \boldsymbol{\mu}_g^{O} + \boldsymbol{\Sigma}_g^{O\bar{O}} (\boldsymbol{\Sigma}_g^{\bar{O}\bar{O}})^{-1} (\mathbf{y}_n^{\bar{O}} - \boldsymbol{\mu}_g^{\bar{O}}),$$

$$\boldsymbol{\Sigma}_g^{O|\bar{O}} = \boldsymbol{\Sigma}_g^{OO} - \boldsymbol{\Sigma}_g^{O\bar{O}} (\boldsymbol{\Sigma}_g^{\bar{O}\bar{O}})^{-1} \boldsymbol{\Sigma}_g^{\bar{O}O}.$$

$\pi_n \left(\boldsymbol{\mu}_{n;g}^{O|\bar{O}}, \boldsymbol{\Sigma}_g^{O|\bar{O}}, \boldsymbol{\Gamma} \right)$ is the conditional joint probability of response pattern $\mathbf{x}_n = (c_{1;n}, \ldots, c_{O;n})$ given the cluster g and the values $\mathbf{y}_n^{\bar{O}}$ for the continuous variables. Finally, τ_g is the probability of belonging to group g subject to $\tau_g > 0$ and $\sum_{g=1}^{G} \tau_g = 1$.

The presence of multidimensional integrals makes the maximum likelihood estimation computationally demanding and infeasible as the number of ordinal variables increases. To overcome this, a composite likelihood approach is adopted [15]. It allows us to simplify the problem by replacing the full likelihood with a surrogate function. As suggested in [21, 23, 24] within a similar context, the full log-likelihood could be replaced by $O(O-1)/2$ marginal distributions each of them composed of a pair of ordinal variables and the $\bar{O}$ continuous variables. In this way, the computational complexity is greatly decreased because the evaluation of the new function requires the calculation of bivariate, rather than O-variate, integrals. This leads to the following surrogate function

$$c\ell(\boldsymbol{\psi}) = \sum_{n=1}^{N}\sum_{i=1}^{O-1}\sum_{j=i+1}^{O}\sum_{c_i=1}^{C_i}\sum_{c_j=1}^{C_j} \delta_{nc_ic_j}^{(ij)} \log\left[\sum_{g=1}^{G} \tau_g \phi_{\bar{O}}(\mathbf{y}_n^{\bar{O}}; \boldsymbol{\mu}_g^{\bar{O}}, \boldsymbol{\Sigma}_g^{\bar{O}\bar{O}}) \pi_{c_ic_j}^{(ij|\bar{O})} (\mu_{n;g}^{(ij|\bar{O})}, \Sigma_g^{(ij|\bar{O})}, \boldsymbol{\Gamma}^{(ij)})\right],$$

where $\delta_{nc_ic_j}^{(ij)}$ is a dummy variable assuming 1 if the nth observation presents the combination of categories c_i and c_j for variables x_i and x_j, respectively, 0 otherwise; $\pi_{c_ic_j}^{(ij|\bar{O})}(\mu_{n;g}^{(ij|\bar{O})}, \Sigma_g^{(ij|\bar{O})}, \boldsymbol{\Gamma}^{(ij)})$ is the conditional probability of the pair $(x_i = c_i, x_j = c_j)$ obtained by integrating the density of a bivariate normal distribution with parameters $(\boldsymbol{\mu}_{n;g}^{(ij|\bar{O})}, \boldsymbol{\Sigma}_g^{(ij|\bar{O})})$ between the corresponding threshold parameters contained in the set $\boldsymbol{\Gamma}^{(ij)}$. The parameter estimates are carried out through an EM-like algorithm that works in the same manner as the standard EM. Likewise, it suffers from the problem of local optima.

In the simulation study, the partition has been initialized randomly. The output of a mixture model for continuous data has been considered as a good rational starting point for the component parameters. On the other hand, the initial values for the thresholds have been computed as follows: for each variable, we have considered the empirical relative frequency of each category and then we have minimized the quadratic difference between this frequency and the corresponding quantile of the mixture.

2.1 *Classification, Model Selection, and Identifiability*

The classification is obtained by assigning the observations to the component with the maximum scaled composite fit, i.e., the CMAP criterion [23, 24]. As regards model selection, the best model is chosen by minimizing the composite version of penalized likelihood selection criteria like BIC or CLC (see [22] and references therein). Finally, as regards identifiability, adopting a composite likelihood approach, the sufficient condition should be reformulated by investigating the Godambe information matrix, that is, the analogous of the information matrix. However, as far as we know, such modification has not been formally investigated yet. About the necessary condition, we note that the number of essential parameters in the block of ordinal variables equals the number of parameters of a log-linear model with only two-factor interaction terms. Thus, it means that we can estimate a lower number of parameters compared to a full maximum likelihood approach. Furthermore, under the underlying response variable approach, the means and the variances of the latent variables are set to 0 and 1, respectively, because they are not identified. This identification constraint individualizes uniquely the mixture components (ignoring the label switching problem), as well described in [19]. This is sufficient to estimate both thresholds and component parameters if all the observed variables have three categories at least and when groups are known. Given the particular structure of the mean vectors and covariance matrices, it is preferable to adopt an alternative, but equivalent, parametrization. This is analogous to that one used by [12]; it consists in setting the first two thresholds to 0 and 1, respectively, without constraining means and variances. This means that there is a one-to-one correspondence between the two sets of parameters. If there is a binary variable, then the variance of the corresponding latent variable is set equal to 1 (while its mean should be still kept free).

3 The Gower Distance Method

Gower distance is computed as the average of partial dissimilarities across observations (subjects or objects), where the computation of the partial dissimilarities depends on the specific type of the variable. For the continuous variables, a range-normalized Manhattan distance is used; for the ordinal variables, they are first ranked, then Manhattan distance is used with a special adjustment for ties. Then, a weighted sum is calculated to create the final distance matrix. However, it is important to note that as the sample size increases, its storage becomes infeasible.

One of the popular partitioning algorithms for mixed-type data is k-medoids (PAM algorithm [13, 25]), which is based on the Gower's distance. The k-means and the PAM algorithm are briefly described in Sects. 3.1 and 3.2. Both suffer from reaching local optima; indeed different initializations can lead to different partitions. Finally, the choice of the number of cluster can be made based on different criteria; the most commonly used is choosing the number of clusters corresponding to an elbow of the scree plot of the within deviance versus the number of clusters.

3.1 k-means

By letting $\mathbf{X} = \{\mathbf{x}_n : n = 1, \ldots, N\}$ be the sample of P-dimensional observations, k-means is based on the minimization of the loss function

$$\ell_{km}(\psi, \mathbf{Z}; \mathbf{X}) = \sum_{n=1}^{N}\sum_{g=1}^{G} z_{ng} d^2(\mathbf{x}_n, \boldsymbol{\mu}_g), \tag{2}$$

where $d^2(\mathbf{x}_n, \boldsymbol{\mu}_g)$ is the squared distance, usually the classical unweighted Euclidean between $\mathbf{x}_n$ and $\boldsymbol{\mu}_g$, $\mathbf{Z} = [z_{ng}]$ is a binary membership matrix, with rows that sum to 1, such that $z_{ng} = 1$ if observation n belongs to cluster g and 0 otherwise, and $\psi = \{\mu_1, \ldots, \mu_G\}$ is the set of cluster centroids.

3.2 k-medoids

The PAM algorithm is an iterative algorithm composed of the following steps:

1. choose k random entities to become the medoids;
2. assign every entity to its closest medoid using the distance matrix computed;
3. for each cluster, the observation with the lowest average distance is re-assigned as the medoid;
4. if at least one medoid has changed, repeat steps 2–4, otherwise the algorithm reaches convergence.

Both k-means and k-medoids are partitioning algorithms and both attempt to minimize the distance between points labeled to be in a cluster and a point designated as the center of that cluster. However, k-means has cluster centers defined by Euclidean distance (i.e., centroids), while cluster centers for PAM are restricted to be the observations themselves (i.e., medoids). Furthermore, k-medoids can be based on an arbitrary dissimilarity matrix. As a consequence, k-medoids is more robust because it minimizes a sum of dissimilarities instead of a sum of squared Euclidean distances.

4 Simulation Study

To evaluate empirically the performance of the different clustering methods, a simulation study has been conducted. We compare: a mixture of Gaussians treating all variables as continuous (Naive), a mixture model for mixed-type data (Mixed), PAM algorithm, and k-means, treating all variables as continuous. The performance has been evaluated in terms of recovering the true cluster structure using the Adjusted Rand Index (ARI) [9] between the true hard partition matrix and the estimated one. The ARI counts the pairs of entities that are assigned to the same or different clusters under both partition matrices. The index has expected value zero for independent clusterings and maximum value 1 for identical clusterings.

We simulated 250 samples from a latent mixture of Gaussians with three components. We considered 8 scenarios given by three different experimental factors: the sample size ($N = 100, 500$), the separation between clusters (well separated or not), and number of ordinal variables (3 ordinal and 5 continuous variables or the other way around).

In order to have approximately the same computational time for each method, the model-based approaches (Naive and Mixed) were initialized using only one good rational starting point described in Sect. 2, while for the remaining ones, 10 different random starting points were used.

Data were generated from a three-component mixture model partially observed with 3 or 5 ordinal variables (5 categories) and 5 or 3 continuous variables. In Table 1, we report the true values that are used to generate the data. The overlap between groups is measured by the Bhattacharyya distance [3, 4]. The Bhattacharyya

Table 1 True values of the observed/latent three-component mixture model and thresholds under different scenarios

Common parameters	
Mixture weights	$p_1 = 0.25$
	$p_2 = 0.35$
	$p_3 = 0.40$
Coviariance matrixes	$\boldsymbol{\Sigma}_1 = \begin{bmatrix} 2.50 & 0.60 & 1.50 & 0.50 & 0.20 & 0.70 & 0.40 & 0.40 \\ 0.60 & 1.00 & 0.40 & 0.40 & 0.65 & 0.40 & 0.50 & 0.20 \\ 1.50 & 0.40 & 2.00 & 0.30 & 0.25 & 0.50 & 0.4 & 0.30 \\ 0.50 & 0.40 & 0.30 & 1.00 & 1.00 & 0.40 & 0.25 & 0.50 \\ 0.20 & 0.65 & 0.25 & 1.00 & 2.00 & 0.70 & 0.65 & 0.20 \\ 0.70 & 0.40 & 0.50 & 0.40 & 0.70 & 1.50 & 0.30 & 0.40 \\ 0.40 & 0.50 & 0.40 & 0.25 & 0.65 & 0.30 & 1.75 & 0.25 \\ 0.40 & 0.20 & 0.30 & 0.50 & 0.20 & 0.40 & 0.25 & 1.00 \end{bmatrix}$ $\boldsymbol{\Sigma}_2 = \begin{bmatrix} 1.875 & 0.450 & 1.125 & 0.375 & 0.150 & 0.5250 & 0.375 & 0.300 \\ 0.450 & 0.750 & 0.300 & 0.300 & 0.4875 & 0.300 & 0.300 & 1.125 \\ 1.125 & 0.300 & 1.500 & 0.225 & 0.1875 & 0.375 & 0.450 & 0.750 \\ 0.375 & 0.300 & 0.225 & 0.750 & 0.750 & 0.300 & 0.5250 & 0.150 \\ 0.150 & 0.4875 & 0.1875 & 0.750 & 1.500 & 0.525 & 0.375 & 0.225 \\ 0.525 & 0.300 & 0.375 & 0.300 & 0.525 & 1.125 & 0.750 & 0.1875 \\ 0.375 & 0.300 & 0.450 & 0.525 & 0.375 & 0.750 & 1.000 & 0.500 \\ 0.300 & 1.125 & 0.750 & 0.150 & 0.225 & 0.1875 & 0.500 & 1.75 \end{bmatrix}$ $\boldsymbol{\Sigma}_3 = \begin{bmatrix} 1.00 & 0.00 & 0.00 & 0.00 & 0.00 & 0.00 & 0.00 & 0.00 \\ 0.00 & 1.00 & 0.00 & 0.00 & 0.00 & 0.00 & 0.00 & 0.00 \\ 0.00 & 0.00 & 1.00 & 0.00 & 0.00 & 0.00 & 0.00 & 0.00 \\ 0.00 & 0.00 & 0.00 & 1.00 & 0.00 & 0.00 & 0.00 & 0.00 \\ 0.00 & 0.00 & 0.00 & 0.00 & 1.00 & 0.00 & 0.00 & 0.00 \\ 0.00 & 0.00 & 0.00 & 0.00 & 0.00 & 1.00 & 0.00 & 0.00 \\ 0.00 & 0.00 & 0.00 & 0.00 & 0.00 & 0.00 & 1.00 & 0.00 \\ 0.00 & 0.00 & 0.00 & 0.00 & 0.00 & 0.00 & 0.00 & 1.00 \end{bmatrix}$
Thresholds	[0, 1, 2, 3]
Separated groups	
Mean Vectors	$\boldsymbol{\mu}_1 = [-1, 3.5, 1.5, 0, -2, 3, 3, 5]$
	$\boldsymbol{\mu}_2 = [2, 0, 4.5, 5, 3, 7, -2, 0]$
	$\boldsymbol{\mu}_3 = [0, -2, -1, -2, 5, -3, 0, -3]$
Non-separated groups	
Mean Vectors	$\boldsymbol{\mu}_1 = [-1, 3.5, 1.5, 0, -2, 3, 0, 5]$
	$\boldsymbol{\mu}_2 = [2, 1, 3, 1.5, 0, 2, -2, 2]$
	$\boldsymbol{\mu}_3 = [0, -1, 0, -0.5, 2, -1, 1.5, -1]$

Table 2 Simulation results: ARI values for different clustering methods across the eight scenarios with $N = 100, 500$, groups with high (H) or low (L) level of separation and number of ordinal variables equal to 3 or 5 with $G = 3$. The Gower distance methods, Gower + PAM (G-PAM) and k-means were initialized using 10 (10) random starting points

3 Ordinal Variable and 5 Continuous Variables				
N = 100 & H	Mixed	Naive	G-PAM (10)	k-means (10)
Mean	0.7997	0.2310	0.5966	0.6566
Median	0.7684	0.1886	0.5947	0.6539
Std	0.1235	0.2209	0.0091	0.0085
N = 500 & H	Mixed	Naive	G-PAM (10)	k-means (10)
Mean	0.9444	0.2618	0.5962	0.6538
Median	0.9663	0.2925	0.5967	0.6544
Std	0.0517	0.1917	0.0064	0.0092
N = 100 & L	Mixed	Naive	G-PAM (10)	k-means (10)
Mean	0.6322	0.1456	0.5824	0.6501
Median	0.6202	0.1066	0.5865	0.6532
Std	0.1096	0.1164	0.0280	0.0121
N = 500 & L	Mixed	Naive	G-PAM (10)	k-means (10)
Mean	0.8953	0.2235	0.5957	0.6543
Median	0.8957	0.1046	0.5962	0.6550
Std	0.0832	0.2416	0.0064	0.0090
5 Ordinal Variable & 3 Continuous Variables				
N = 100 & H	Mixed	Naive	G-PAM (10)	k-means (10)
Mean	0.6895	0.2223	0.5921	0.6125
Median	0.6354	0.1437	0.5891	0.6095
Std	0.1547	0.2271	0.0124	0.0124
N = 500 & H	Mixed	Naive	G-PAM (10)	k-means (10)
Mean	0.8181	0.3725	0.5898	0.6124
Median	0.8435	0.3511	0.5882	0.6089
Std	0.1096	0.2735	0.0088	0.0151
N = 100 & L	Mixed	Naive	G-PAM (10)	k-means (10)
Mean	0.6073	0.1080	0.5545	0.6458
Median	0.5634	0.0113	0.5553	0.6438
Std	0.1321	0.1877	0.0254	0.0120
N = 500 & L	Mixed	Naive	G-PAM (10)	k-means (10)
Mean	0.8069	0.2027	0.5454	0.6432
Median	0.8150	0.1255	0.5413	0.6423
Std	0.1442	0.2342	0.0130	0.0080

distance is equal to: 19.00 considering $g = 1, 2$, 26.27 considering $g = 1, 3$ and 34.27 considering $g = 2, 3$ when the groups are well separated; 5.96 considering $g = 1, 2$, 12.98 considering $g = 1, 3$ and 11.24 considering $g = 2, 3$ when the groups are not well separated. In the simulation study, the number of groups is kept fixed. Indeed, the purpose of the study is to assess the ability of the algorithm to capture the cluster structure. In Table 2 we report the simulation results.

Analyzing the results in Table 2, we note that all clustering methods improve their performances as N increases and the level of separation between groups is higher, as expected. In almost all scenarios, the mixture model for mixed-type data seems to behave better than others. Indeed, we note that in terms of mean or median the mixture model for mixed-type data is the best, followed by the k-means and PAM based on the Gower distance matrix. The poorest performances are shown by the naive approach. In terms of mean or median, the mixture model for mixed-type data is not always the best compared to the non-model-based approaches. More specifically, when $N = 100$ and the groups are not well separated, it seems that it is more affected by the issue of local maxima. Furthermore, we note that when there are more ordinal variables than continuous variables, ARI values decrease, although when N increases the worsening is not significant. This is expected, since more ordinal variables we have, more information is losing about the cluster structure underlying the data. Finally, although it is still common to treat ordinal data as metric, we have shown that it can lead to wrong results, especially when the groups are not well separated.

5 Concluding Remarks

In this paper, we compared the model-based approach and Gower distance methods to cluster mixed-type data. From the simulation study, it is possible to conclude that when the groups are less separated, the clustering performances of the Gower distance methods seem to be more affected by the choice of the random starting points. The model-based for mixed type of data as N increases becomes the best one both in terms of means and median. However, it is important to note that larger sample sizes could cause some computational problems. On one hand, for larger N it is possible to compute the Gower matrix, but its storage may become infeasible. On the other hand, this leads to a higher number of bivariate integrals involved in the composite likelihood. However, this increase remains linear, and thus still feasible.

References

1. Agresti, A.: Analysis of Ordinal Categorical Data, vol. 656. Wiley (2010)
2. Ahmad, A., Khan, S.S.: Survey of state-of-the-art mixed data clustering algorithms. IEEE Access **7**, 31883–31902 (2019)

3. Bagnato, L., Greselin, F., Punzo, A.: On the spectral decomposition in normal discriminant analysis. Commun. Stat. - Simul. Comput. **43**(6), 1471–1489 (2014)
4. Bhattacharyya, A.: On a measure of divergence between two multinomial populations. Sankhya: Ind. J. Stat. (1933-1960) **7**(4), 401–406 (1946)
5. Everitt, B.: A finite mixture model for the clustering of mixed-mode data. Stat. Prob. Lett. **6**(5), 305–309 (1988)
6. Foss, A.H., Markatou, M., Ray, B.: Distance metrics and clustering methods for mixed-type data. Int. Stat. Rev. **87**(1), 80–109 (2019)
7. Gollini, I., Murphy, T.: Mixture of latent trait analyzers for model-based clustering of categorical data. Stat. Comput. **24**(4), 569–588 (2014)
8. Gower, J.C.: A general coefficient of similarity and some of its properties. Biometrics **27**(4), 857–871 (1971)
9. Hubert, L., Arabie, P.: Comparing partitions. J. Classif. **2**(1), 193–218 (1985)
10. Hunt, L., Jorgensen, M.: Clustering mixed data. WIREs Data Min. Knowl. Disc. **1**(4), 352–361 (2011)
11. Jöreskog, K.G.: New developments in lisrel: analysis of ordinal variables using polychoric correlations and weighted least squares. Quality and Quantity **24**(4), 387–404 (1990)
12. Jöreskog, K.G., Sörbom, D.: LISREL 8: User's Reference Guide. Scientific Software (1996)
13. Kaufman, L., Rousseeuw, P.J.: Clustering by means of medoids (1987)
14. Lee, S.Y., Poon, W.Y., Bentler, P.: Full maximum likelihood analysis of structural equation models with polytomous variables. Stat. Prob. Lett. **9**(1), 91–97 (1990)
15. Lindsay, B.: Composite likelihood methods. Contemp. Math. **80**, 221–239 (1988)
16. Lubke, G., Neale, M.: Distinguishing between latent classes and continuous factors with categorical outcomes: Class invariance of parameters of factor mixture models. Multivariate Behav. Res. **43**(4), 592–620 (2008)
17. McLachlan, G., Peel, D.: Finite Mixture Models. Wiley (2000)
18. Mechelen, I., Boulesteix, A., Dangl, R., Dean, N., Guyon, I., Hennig, C., Leisch, F., Steinley, D.: Benchmarking in cluster analysis: A white paper. arXiv: Other Statistics (2018)
19. Millsap, R.E., Yun-Tein, J.: Assessing factorial invariance in ordered-categorical measures. Multivariate Behav. Res. **39**(3), 479–515 (2004)
20. Muthén, B.: A general structural equation model with dichotomous, ordered categorical, and continuous latent variable indicators. Psychometrika **49**(1), 115–132 (1984)
21. Ranalli, M., Rocci, R.: Mixture models for ordinal data: a pairwise likelihood approach. Stat. Comput. 1–19 (2016). https://doi.org/10.1007/s11222-014-9543-4
22. Ranalli, M., Rocci, R.: Standard and novel model selection criteria in the pairwise likelihood estimation of a mixture model for ordinal data. In: Adalbert, F.X., Hans, W., Kestler, A. (eds.) Analysis of Large and Complex Data. Studies in Classification,Data Analysis and Knowledge Organization (2016). https://doi.org/10.1007/978-3-319-25226-1
23. Ranalli, M., Rocci, R.: Mixture models for mixed-type data through a composite likelihood approach. Comput. Stat. Data Anal. **110**(C), 87–102 (2017). https://doi.org/10.1016/j.csda.2016.12.01
24. Ranalli, M., Rocci, R.: A model-based approach to simultaneous clustering and dimensional reduction of ordinal data. Psychometrika (2017). http://orcid.org/10.1007/s11336-017-9578-5
25. Steinley, D.: Handbook of Cluster Analysis, chap. *K*-Medoids and Other Criteria for Crisp Clustering. Chapman and Hall/CRC, New York (2016)

Exploring the Gender Gap in Erasmus Student Mobility Flows

Marialuisa Restaino, Ilaria Primerano, and Maria Prosperina Vitale

Abstract The present contribution aims at exploring the Erasmus student mobility flows across European countries given the relevant role played by the internationalisation process in the implementation of university policies. In particular, the main purpose is to confirm the presence of a gender gap across countries in the Erasmus programme according to the related literature. Mobility data and socio-demographic indicators are collected from the European Union Open Data Portal and the Eurostat website. Information on student flows are then considered to define network data structures in which the nodes are the countries and the incoming and outgoing links represent the number of students exchanged between countries. Results show that the number of females involved in Erasmus programme is greater than the number of males, even if the position of countries in terms of centrality scores in the network structure remains similar.

Keywords Gender gap · Erasmus student mobility · European open data · Network measures · Clustering

1 Introduction

The internationalisation could be defined as "the process of integrating an international, intercultural or global dimension into the purpose, functions or delivery of post-secondary education" [12]. Among others, the degree of internationalisation in higher education is measured by the reception of foreign students and the sending of students abroad. In fact, universities consider the number of foreign students

M. Restaino (✉) · I. Primerano · M. P. Vitale
University of Salerno, Fisciano, Italy
e-mail: mlrestaino@unisa.it

I. Primerano
e-mail: iprimerano@unisa.it

M. P. Vitale
e-mail: mvitale@unisa.it

S. Balzano et al. (eds.), *Statistical Learning and Modeling in Data Analysis*, Studies in Classification, Data Analysis, and Knowledge Organization,
https://doi.org/10.1007/978-3-030-69944-4_18

they attract as an indicator of the attractiveness and the reputation of their education provisions.

The most famous mobility programme developed by the European Union (EU) to promote the exchange of cultural, professional and personal experiences within EU countries is the European Region Action Scheme for the Mobility of University Student, that is the Erasmus programme. The participation in this programme has increased from 3,000 participants in 1987 to 272,497 in 2013–2014, and within the new Erasmus+ for the period 2014–2020, the number of participants has increased to 796,761 for Key Action 1 in 2017.

The benefits of participating in study abroad programme are mainly related to the personal and professional growth of students. The development of learning experience with intercultural and linguistic improvement skills and the enhancement of job prospects and opportunities after graduation are the main factors explored for students involving in this international experience [1, 11, 13–15].

Within this scenario, the analysis of how gender might relate to the international student mobility trajectories is taken up by some authors, showing as female students are often overrepresented in Erasmus [2, 5]. This tendency of a "strong gender bias in favour of female students" is discussed in the recent contribution of De Benedictis and Leoni [see [9] and references therein].

The present contribution aims at analysing the gender (in)equality in Erasmus mobility by investigating if there exists any differences in incoming–outgoing flows of students between European countries in six academic years, from 2008–2009 to 2013–2014. To capture the structural features and patterns of Erasmus mobility flows by gender, the adoption of network measures [3, 4, 8] along with clustering techniques is able to identify groups of good importers and good exporters countries involved in this process.

The data under study are gathered from the European Union Open Data Portal, and network data structures are defined in order to analyse and describe relationships among countries. Moreover, educational indicators are collected from the Eurostat website to describe the investments of European countries in higher education in the period under analysis and to better clarify the role of each country in the internationalisation process of higher education system.

The contribution is organised as follows. Section 2 briefly describes the data and the methodological approach for exploring international student mobility data and country indicators. Section 3 reports the main findings and some suggestions for further developments.

2 Data and Methods

The data on Erasmus student mobility flows are downloaded by the official European Commission website on Erasmus-Statistics[1] for six academic years, from 2008–2009

[1]For details see https://data.europa.eu/euodp/en/data/publisher/eac.

Table 1 List of indicators gathered from the Eurostat website

Index
Enrolment in tertiary education
Expenditure on tertiary as a percentage of government expenditure on education
Net flow ratio of internationally mobile students (inbound - outbound)
School-age population, tertiary education
Government expenditure on education as a percentage of GDP
Gross enrolment ratio, tertiary education
Graduates from tertiary education
Expenditure on education as a percentage of total government expenditure
Government expenditure on tertiary education as a percentage of GDP
Teachers in tertiary education programmes
Inbound mobility rate
Expenditure on tertiary as a percentage of total government expenditure
Total outbound internationally mobile tertiary students studying abroad
Total inbound internationally mobile students
Outbound mobility ratio
Gross outbound enrolment ratio

to 2013–2014. Two types of Erasmus mobility of students enrolled at higher education institutions are collected: the *Student Mobility for Studies* (SMS) that enables students to spend a study period in another country, and the *Student Mobility for Placement* (SMP) that enables students to spend a placement period (traineeship or internship) in an enterprise/organisation in another country. The information available in the datasets are ID of sending and hosting Partner Erasmus; sending and hosting countries; students' gender; subject area code; type of mobility (SMS or SMP); level of study (first cycle, second cycle, third cycle and short cycle); duration of mobility in months.

The Erasmus data are used to defined network structures over time represented by a weighted digraph $\mathcal{G}(\mathcal{V}, \mathcal{L}, \mathcal{W})$, where $\mathcal{V}$ is the set of countries (vertices), $\mathcal{L} \subseteq \mathcal{V} \times \mathcal{V}$ is the set of arcs (directed lines) and $\mathcal{W}$ is the set of weights, $w : \mathcal{L} \rightarrow \Re$, i.e. the number of students exchanged between pairs of countries. The corresponding adjacency matrix $\mathbf{A}$ is both not symmetric, with a directed link from the origin country to the destination country, and weighted, with elements $a_{ij} = w(v_i, v_j) = w_{ij}$ greater than 0 if there is a link between country v_i and country v_j, and $a_{ij} = 0$ otherwise.

In addition, to inspect the attractiveness of universities, several indicators downloaded from the Eurostat website and related to specific features of the Tertiary Education System[2] are added as further information in the analysis (Table 1).

Social Network Analysis (SNA) tools and exploratory data analysis methods are then considered as a strategy of analysis to capture the structural characteristics and

[2]For details see https://ec.europa.eu/eurostat/statistics-explained/index.php.

patterns of student mobility flows in the Erasmus programme in order to confirm whether a gender gap exists. First, to study the temporal changes and the networks' characteristics for the six academic years under analysis, weighted directed adjacency matrices are defined. Each matrix describes the student's flows among countries involved in the Erasmus programme for each academic year by type of Erasmus programme and by gender. Then, to identify countries who play a central role, the hub and authority centrality scores [10] are adopted to determine which countries are good exporters (i.e. countries with good hub points to many other countries) and/or good importers (i.e. countries with a high authority score is linked by many different hubs). The peculiar structure of student mobility flows by gender is considered to discover potential differences in the Erasmus country destinations of males and females. Second, the network results are enriched by considering exploratory data analysis methods (i.e. principal component analysis and hierarchical clustering) applied to both higher educational indicators and network measures, to reveal connections between the roles played by countries in the student mobility network and their investments in education as a key element of institutions' attractiveness.

In Sect. 3, we report the main findings showing the trend of the Erasmus mobility, and the temporal changes and the networks' characteristics to underline the differences in Erasmus country destinations of males and females.[3]

3 Results

The Erasmus mobility networks have mainly changed in terms of number of students involved in the programme over time. In general, the number of males and females students who joined the Erasmus programme increased. A remarkable difference between the networks of SMS and SMP for males and females is observed. The number of students who moved for study is greater than the number of students who moved for placement. Moreover, the number of females who go abroad for study and for placement is greater than that of men. These results are shown in Table 2, where the distribution of Erasmus students by gender for SMS and SMP and over time is displayed.

In particular, the number of students goes up from 168,193 in 2008–2009 to 212,208 in 2013–2014 for the SMS network (+26.2%) and from 30,330 in 2008–2009 to 60,289 in 2013–2014 for the SMP (+98.8%) (Table 2). The number of women increases from 101,982 in 2008–2009 to 127,782 in 2013–2014 in SMS (+25.3%) and from 18,609 in 2008–2009 to 37,107 in 2013–2014 for the SMP (+99.4%). Then, the number of men becomes larger from 66,211 in 2008–2009 to 84,426 in 2013–2014 in SMS (+27.5%) and from 11,721 in 2008–2009 to 23,182 in 2013–2014 for the SMP (+97.8%) (Table 3).

[3]The analysis is performed by the open-source R packages "sna", "igraph" and "blockmodeling" [6, 7, 16].

Table 2 Distribution of Erasmus Student Mobility networks for Studies (SMS) and for Placement (SMP) by gender from 2008–2009 to 2013–2014

Year	Total number of exchanges	#. of exchanges		% of females	
		SMS	SMP	SMS	SMP
2008–2009	198,523	168,193	30,330	60.6	61.4
2009–2010	213,266	177,705	35,561	61.1	60.9
2010–2011	231,408	190,495	40,913	60.9	61.8
2011–2012	252,827	204,744	48,083	60.6	61.1
2012–2013	268,143	212,522	55,621	60.6	61.9
2013–2014	272,497	212,208	60,289	60.2	61.6

Our elaboration based on *Erasmus Facts, Figures and Trends*, European Commission website

The structure of the temporal networks shows in the six academic years under analysis a little increase in terms of involved countries and links among them (Table 3). Specifically, the number of countries for the SMS and SMP networks increases from 31 in 2008–2009 to 33 and 34 in 2013–2014 for males and females. Moreover, for females the number of links goes up from 769 links in 2008–2009 to 896 links in 2013–2014 in the SMS network and from 591 links in 2008–2009 to 796 links in 2013–2014 for the SMP network. Then, for males the number of links goes up from 760 links in 2008–2009 to 874 links in 2013–2014 in the SMS network and from 569 links in 2008–2009 to 761 links in 2013–2014 for the SMP network.

Looking at the number of outgoing and incoming students, countries are classified as good exporters and/or good importers by means of the hub and authority network centrality indexes. These classifications are drawn up for males and females and for SMS and SMP. We note that the ranking for SMS network is stable across the years. In particular, Spain, France, United Kingdom, Italy and Germany are always the most favourite destinations. The other five positions (from 6 to 10) show a little change between males and females. For example, Denmark is a destination chosen by men in the first two years. The women prefer Belgium. This difference should be related to the fields of study. In fact, looking at the raw data for these two countries, it emerges that the males studied Economics and Engineering in Denmark, while the females went to Belgium to study Political Sciences, Foreign Languages and Health. This result is in line with those showed in [9], where the authors analysed the gender bias in Erasmus mobility by looking at the fields of study. For SMP network the ranking is stable over the period, also if we look at the gender level.

Then, for the SMS network, the best importing countries obtaining high values for authority scores are Spain, France and United Kingdom; while Germany and France show the highest hub scores. Looking at the gender level, it emerges that the best importing countries are different between females and males. In particular, for females Spain and France are always in the first two positions for all years, while for males only Spain confirms its first position, and the second position changes over the time. For almost all the years, Spain is both the best authority and hub country in the SMS network. In both rankings, Italy is always in the top five positions. In particular,

Table 3 Number of countries involved in the Erasmus Student Mobility networks for Studies (SMS) and for Placement (SMP) by gender from 2008–2009 to 2013–2014

	Female						Male					
Academic	Erasmus SMS			Erasmus SMP			Erasmus SMS			Erasmus SMP		
years	Countries	Links	Students	Countries	Links	Students	Countries	Links	Students	Countries	Links	Students
2008–2009	31	769	101,982	31	591	18,609	31	760	66,211	31	569	11,721
2009–2010	32	713	108,148	31	651	22,082	32	761	69,557	31	597	13,479
2010–2011	33	804	115,934	33	665	25,267	33	785	74,561	33	616	15,646
2011–2012	33	884	124,103	33	740	29,365	33	866	80,641	33	712	18,718
2012–2013	33	882	128,562	33	777	34,399	33	858	83,433	34	740	21,150
2013–2014	34	896	127,782	33	796	37,107	34	874	84,426	34	761	23,182

it is in a better position in the hub score rankings, showing a better exporting than importing behavior in the SMS network. Poland appears to be a good exporting country for both females and males. Moreover, in the last three years also Turkey enters in the top five positions.

Furthermore, for the SMP network, the rankings of countries obtained by the authorities scores show that the best importing countries for Erasmus placement are United Kingdom and Spain. Looking at the ranking with respect to gender, we see that the best two importing countries for females are United Kingdom and Spain, while for males the first two importing countries change in 2011–2012. In fact, United Kingdom is replaced by Germany. At the same time, the best exporting countries are France and Germany, showing the highest values of the hubs score. Italy has a marginal role in the SMP network, since it is between the fifth and -sixth position. Considering the ranking for females, we note that the best three exporting countries are France, Germany and United Kingdom, even if the order changes over the period considered. The three best countries for males are Denmark, Germany and Spain, except in last year when Italy ascends the ranking getting the third position.

To better describe the structure of student mobility flows, Principal Component Analysis (PCA) and Hierarchical Clustering are performed on Erasmus data collected in the academic year 2013–14 by considering the hub and authority centrality measures and some indicators of Tertiary Education System described in Table 3. The analysis considers separately the type of Erasmus programme and the gender. Starting from the PCA results,[4] the agglomerative hierarchical clustering with Ward's criterion is performed to identify the presence of groups of countries. For all cases, *three clusters* have been identified (see Figs. 1 and 2).

The 34 countries joining the Erasmus programme for studies and for placement, considering males, are grouped as follows:

- *cluster 1* (7 countries): Germany, Spain, France, Italy, Poland, Turkey and the United Kingdom;
- *cluster 2* (24 countries): Austria, Belgium, Bulgaria, Czech Republic, Denmark, Estonia, Finland, Greece, Croatia, Hungary, Ireland, Iceland, Lithuania, Latvia, Macedonia, Malta, The Netherlands, Norway, Portugal, Romania, Sweden, Slovenia, Slovakia and Switzerland;
- *cluster 3* (3 countries): Cyprus, Liechtenstein and Luxembourg.

As for the females, there is a difference in the first and second cluster. For both -SMS and -SMP networks, Cluster 1 is made up of 6 countries, while cluster 2 of 25 countries. The country moving from cluster 1 to cluster 2 is Poland. However, the countries in *cluster 1* are the most central ones in the SMS and SMP networks for males and females, showing the highest hub and authority scores. The countries in *cluster 3* are the less central ones in the networks, showing the lowest scores. In *cluster 2* there are the less influential countries in the Erasmus programme, with hub and authority scores closer to 0.

---

[4]The results of PCA are available upon request.

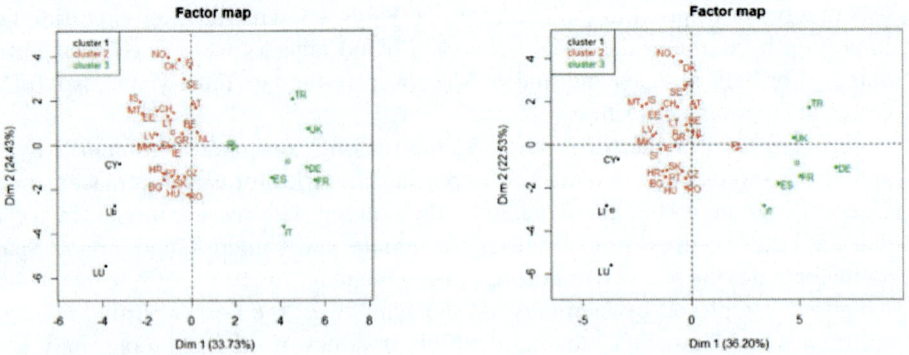

(a) 2013-2014 - SMS - Female (b) 2013-2014 - SMS - Male

Fig. 1 Factorial map of the first two principal components on educational indicators in Erasmus student mobility for studies (SMS) for females and males. Countries are coloured according to the three clusters' solutions. AT = Austria; BE = Belgium; BG = Bulgaria; CH = Switzerland; CY = Cyprus; CZ = Czech Republic; DE = Germany; DK=Denmark; EE = Estonia; ES = Spain; FI = Finland; FR = France; GR = Greece; HR = Croatia; HU = Hungary; IE = Ireland; IS = Iceland; IT = Italy; LI = Liechtenstein; LT = Lithuania; LU = Luxembourg; LV = Latvia; MK = Macedonia; MT = Malta; NL = Netherlands; NO = Norway; PL = Poland; PT = Portugal; RO = Romania; SE = Sweden; SI = Slovenia; SK = Slovakia; TR = Turkey; UK = United Kingdom

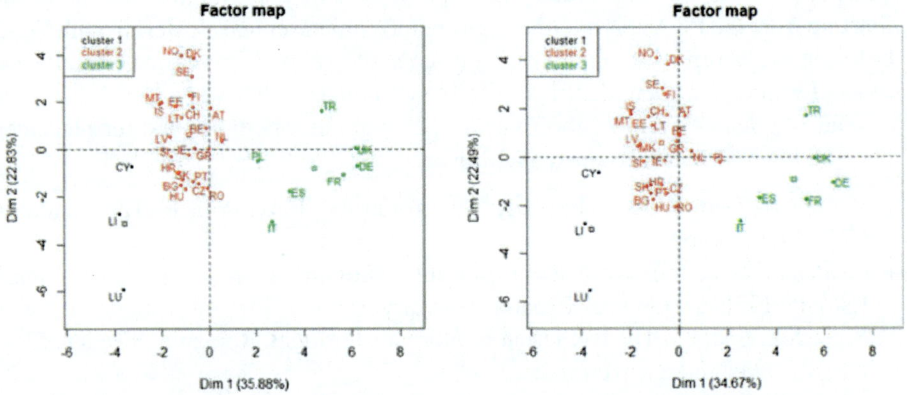

(a) 2013-2014 - SMP - Female (b) 2013-2014 - SMP - Male

Fig. 2 Factorial map of the first two principal components on educational indicators in Erasmus Student Mobility for Studies (SMS) and for Placement (SMP) for females and males. Countries are coloured according to the three clusters' solutions. AT = Austria; BE = Belgium; BG = Bulgaria; CH = Switzerland; CY = Cyprus; CZ = Czech Republic; DE = Germany; DK = Denmark; EE=Estonia; ES = Spain; FI = Finland; FR = France; GR = Greece; HR = Croatia; HU = Hungary; IE = Ireland; IS = Iceland; IT = Italy; LI = Liechtenstein; LT = Lithuania; LU = Luxembourg; LV = Latvia; MK = Macedonia; MT = Malta; NL = Netherlands; NO = Norway; PL = Poland; PT = Portugal; RO = Romania; SE = Sweden; SI = Slovenia; SK = Slovakia; TR = Turkey; UK = United Kingdom

Summarising, even if the number of all students who joined the Erasmus programme increased from 2008–2009 to 2013–2014, we note that the number of females involved in the SMS and -SMP is greater than the number of males. This result is in line with the results reported in related literature. As a justification of this gender bias persisting over time across countries, we can consider the effect of the fields of study as discussed in De Benedictis and Leoni [9]. The authors using the same data of the EU open data portal but at university level justify the advantage of female participation over male in this programme given the denser network of connections involving female students. These latter prevail in fields such as Arts and Humanities, Education and Social Sciences, Journalism and Information; whereas the bias in favour of female students is strongly reduced in fields such as Information and Communication Technologies and Engineering, Manufacturing and Construction. The position of countries according to the hub and authority scores for SMS and SMP, instead, is similar at gender level.

As further lines of research, we are interested in analysing the configuration of Erasmus student network over time with respect to the attractiveness of each country to better investigate the gender gap in the internationalisation process, by adding some information on the tourism behavior in the European countries, such as number of trips, overnight stays, and the values for travel expenditures.

References

1. Amendola, A., Restaino, M.: An evaluation study on students international mobility experience. Qual. Quant. **51**(2), 525–544 (2017)
2. Böttcher, L., Araújo, N.A.M., Nagler, J., Mendes, J.F.F., Helbing, D., Herrmann, H.J.: Gender gap in the ERASMUS mobility program. PLoS One **11**(2), (2016)
3. Breznik, K., Skrbinjek, V, Law, K., Dakovic, G.: On the Erasmus Student Mobility for Studies. In: Dermol, V., Trunk Sirca, N., Dakovic, G. (eds.). Active citizenship by knowledge management & innovation: proceedings of the Management, Knowledge and Learning International Conference 2013, 19-21 June 2013, Zadar, Croatia, (MakeLearn, ISSN 2232-3309). Bangkok; Celje; pp. 1371–1377. ToKnowPress, Lublin (2013)
4. Breznik, K., Ragozini, G.: Exploring the Italian erasmus agreements by a network analysis perspective. In: Proceeding of the 2015 IEEE/ACM International Conference on Advances in Social Networks Analysis and Mining (ASONAM), pp. 837–838. Paris (2015)
5. Cairns, D.: Researching social inclusion in student mobility: methodological strategies in studying the Erasmus programme. IOSR-JRME **42**(2), 137–147 (2019)
6. Carter T. B.: sna: Tools for Social Network Analysis. R package version 2.5. (2019). https://CRAN.R-project.org/package=sna
7. Csardi G., Nepusz T.: The igraph software package for complex network research. Int. J. Compl. Syst. **1695**. (2006) http://igraph.org
8. Derszi, A., Derszy, N., Kaptalan, E., Neda, Z.: Topology of the Erasmus student mobility network. Phys. A **390**(13), 2601–2610 (2011)
9. De Benedictis, L., Leoni, S.: Gender bias in the Erasmus students network (2020). arXiv:2003.09167
10. Kleinberg, J.M.: Authoritative sources in a hyperlinked environment. J. ACM **46**(5), 604–632 (1999)

11. King, R., Ruiz-Gelices, E.: International student migration and the european year abroad: effects on european identity and subsequent migration behaviour. Int. J. Popul. Geogr. **9**(3), 229–252 (2003)
12. Knight, J.: Internationalization: concepts, complexities and challenges. In: Forest, J.J., Altbach, P.G. (eds.) International Handbook of Higher Education, pp. 207–227. Springer, Dordrecht (2007)
13. Norris, E.M., Gillespie, J.: How study abroad shapes global careers: evidence from the United States. J. Stud. Int. Edu. **13**, 392–397 (2009)
14. Parey, M., Waldinger, F.: Studying abroad and the effect on international labour market mobility: evidence from the introduction of Erasmus. Econ. J. **121**(551), 194–222 (2011)
15. Restaino, M., Primerano, I., Vitale, M.P.: Analysing international student mobility flows in higher education: a comparative study on european countries. Soc. Ind. Res. (2020). https://doi.org/10.1007/s11205-020-02282-2
16. Ziberna, A.: Generalized and Classical Blockmodeling of Valued Networks, R package version 0.3.4. (2018)